# Numanities - Arts and Humanities in Progress

Volume 26

The series originates from the need to create a more proactive platform in the form of monographs and edited volumes in thematic collections, to contribute to the new emerging fields within art and humanistic research, and also to discuss the ongoing crisis of the humanities and its possible solutions, in a spirit that should be both critical and self-critical.

"Numanities" (New Humanities) aim at unifying the various approaches and potentials of arts and humanities in the context, dynamics and problems of current societies. The series, indexed in Scopus, is intended to target a broad academic audience. Aside from taking interest in work generally deemed as 'traditional humanities research', Numanities are also focused on texts which meet the demands of societal changes. Such texts include multi/inter/cross/transdisciplinary dialogues between humanities and social, or natural sciences. Moreover, the series is interested also in what one may call "humanities in disguise", that is, works that may currently belong to non-humanistic areas, but remain epistemologically rooted in a humanistic vision of the world. We also welcome are less academically-conventional forms of research animated by creative and innovative humanities-based approaches, as well as applied humanities. Lastly, this book series is interested in forms of investigations in which the humanities monitor and critically asses their scientific status and social condition.

This series will publish monographs, edited volumes, and commented translations.

Simona Stano

# Critique of Pure Nature

Springer

Simona Stano
University of Turin (UniTO)
Turin, Italy

ISSN 2510-442X ISSN 2510-4438 (electronic)
Numanities - Arts and Humanities in Progress
ISBN 978-3-031-45074-7 ISBN 978-3-031-45075-4 (eBook)
https://doi.org/10.1007/978-3-031-45075-4



This Springer imprint is published by the registered company Springer Nature Switzerland AG
The registered company address is: Gewerbestrasse 11, 6330 Cham, Switzerland

Paper in this product is recyclable.

*To Dafne,*
*who has shown me*
*the deepest meaning of everything*

# Foreword

## Nature, Naturality, Naturalness

Many languages reveal the need to distinguish between two different relations to nature by referring to it with as many related but disparate verbal expressions. Naturality is not nature. Indeed, nature is an unattainable asymptote of any semiosis for which a total adherence to the wild state of reality is evoked, untouched by the determinations of language. There is perhaps no better way to understand the distinction between nature and naturality than to think about the disciplines and thus the networks of thoughts, knowledge and languages that deal with them. The so-called natural sciences do not deal with naturality. They lean into nature. That is, they try to grasp it as it is, and not as it seems, in its being and making, and not in its seeming and appearing. Above all, they are totally disinterested in its supposed having to be, in its deontology. Nature is, and that is enough. However, as the semiotics of the scientific discourse, which has set out to deconstruct the rhetoric of science, well knows, one cannot talk about nature. One can perhaps measure it, quantify it, count it, but the moment one commensurates it, qualifies it, and above all, recounts it, it becomes other than itself, it becomes, that is, a nature translated but also misrepresented by language, by semiosis: an interpreted nature.

Sharpening the gaze of analysis and criticism, then, we have come to realise that nature is semiotised not only in the narrative of scientific dissemination, but also in the very practices supposedly aimed at capturing it in its nakedness. Indeed, semiotics is even more ambitious than linguistics in its attempt to unearth configurations of meaning even beyond verbal language. That water is transcribable as $H_2O$ is certainly a fact of nature, but that this metalanguage and not another was chosen to describe and annotate its internal structure is the result of a stylistic choice as well. Nature, then, exists as subsistence of network of causes and effects that are beyond meaning, but as soon as this network interacts with the human, it detaches itself from the pure objectivity of forces and agencies and unravels, instead, a chain of signs, many of which Peirce's semiotics calls indexical to point out precisely that they are rooted in reality but distinguished from it by the intervention of that mysterious but foundational process of humanity that is semiosis.

It seemed natural to the German astronomer Simon Marius, who wanted to name the satellites of Jupiter, to do so after the names of Zeus' mistresses; the following scientific literature adopted, it is true, a much more algid system of ordinal numerals, but in 1975, the International Astronomical Union formed a working group, the 'Task Group for Outer Solar System Nomenclature', to return to naming in the style of Marius, so that we went back to the god's mistresses and, since 2004, having finished with the albeit very numerous mistresses, began with the even more numerous descendants. The anecdote makes one smile, but it can be used to emphasise how, even in the natural, exact and hard sciences, or whatever else you want to call them, numbering nature is not enough; even scientists feel a very strong desire to move from numbers to names, and to speak about nature in the same language with which human beings speak and talk to each other, and, as semioticians know well, are also 'spoken' by nature.

Indeed, language is, at least in its evolutionary both phylogenetic and ontogenetic genesis, a 'fact of nature', yet extraordinarily it turns on itself, speaking of itself, of the nature from which it emerged, often muddying the waters, disguising or simulating its origins and staging fictitious relationships of forces and hierarchies between nature and language. Language (or to be more semiotically general, 'semiosis') emerges from a natural substratum, but then pretends to emancipate itself from it, attributing to itself the same naked reality of nature, presenting the meaning it constructs as being able to find, in enunciation and its products, in the discourse that speaks of nature as well as in the texts it brings about, a nature linguistically reflected, semiotically mirrored, adamantine as the object that it means.

As soon as language touches nature, however, it undoes it. The débrayage of language in relation to nature, that is, its emergence from a natural substratum to overturn itself and its own naturing nature, immediately gives rise to a 'meta' level that is distinct but also distant from nature itself. When nature makes sense to us humans, that is, always, except when we are completely unaware and unconscious of it, it also makes itself partly language. Naturality, then, is essentially an embrayage, the term French structuralist semiotics adopts to refer to those processes of meaning in which a return of language to the unattainable origin of its enunciation is simulated. It, naturality, effuses itself as an effect of meaning generated by any practice that poses and proposes itself as capable of guaranteeing to the human an ascent of the flow of language back to its natural source. It promises, that is, to flatten that 'meta' level that inevitably occurs in semiosis, as if it were possible to conflate it with its primal substratum, and make it possible, therefore, that in nature there is directly meaning without language, that is, according to a 'natural' language, and at the same time that in meaningful language there is immediately nature.

These attempts at embrayage, which are and can only be rhetorical constructions, have a long history, beginning with Pliny's attempt to write a 'natural history' of humanity, for example, but also with the very meaning of 'natural languages', which also has a remote genealogy, whereon we cannot dwell here. In short, and in summary, however, what is a 'natural' language, if not a language suggested to have had a genesis and evolution 'in contact with nature', adhering as it were to its lines of development, to the aforementioned network of causes and effects, and thus an evolution distinct from that of 'artificial' or rather 'planned' languages?

But already the fact that we cannot actually define exactly what a 'natural language' is except by contrasting it with an 'artificial language' should be suspicious. New 'invented' languages were needed so that the already existing ones might be adorned not with a mantle of 'nature' but of 'naturality'; the rhetoric of natural languages is also a rhetoric, and as such it responds exactly to the aforementioned dynamics of embrayage. On closer inspection, in fact, one realises that many planned, invented, or 'artificial' languages evolve with traits, according to dynamics, and with characteristics germane to those of the so-called 'natural' languages; and above all, one realises, in this comparison, that many of the elements that prompt one to designate a language as 'artificial', and thus as departing from the 'natural' ones, are also found in the genesis and development of the latter. One of them is certainly central: individual or circumscribed group intentionality in constructing a language. But can it be said that this element, individual intentionality, plays no role whatsoever in the giving and making of 'natural languages'? Do not writers change natural language, and with it the nature of language? Do we not also do so a bit ourselves, day by day, through our own speech?

Removing intentionality, or even agency, from the scene, the dialectic between langue and parole as postulated by Saussure and deepened by Benveniste would have no subsistence or meaning. We humans construct our 'natural languages' day after day, albeit according to measures of scale often far greater than those implicit in 'artificial' language formations. And after all, then, if languages were really the product of a neutral and unintentional agency, of a pure collective intelligence, why would we have so many plural 'natural languages' and not just one? Instead, we must again reiterate that languages are natural in the sense of naturality, and not in that of nature, and that conversely languages are artificial in the sense of artificiality, and not necessarily in that of artifice. This form of thinking becomes increasingly useful and urgent today, as we are developing forms of artificial intelligence whose processes more and more resemble those which, in the development of natural intelligence, have been imputed, precisely, to an unintentional matrix, to an emergent agency. It is possible that, in the long run, even artificial intelligence, beyond a certain level of internal complexity, will begin to enunciate its own naturality. But even then, it will be necessary to do the work of semioticians, and to remember how much artificiality there is in our own intelligence, or how much it owes to the linguistic, semiotic and cultural crucible in which it coagulates.

It is precisely in relation to this panorama of very broad horizons that the 'critique' proposed by Simona Stano has the taste and boldness of the philosophical gestures of other epochs, those that set out to discuss very general systems and to unveil deep laws of human action. Well, if one were to summarise in one line the entire book that is humbly presented here—a complex book, the fruit of research and study over many years, of incessant reworking and extraordinary concentration around a very clear thread, but unravelled through a very wide range of interests—indeed, if one were to summarise or at least announce this which is evidently the fruit of the indefatigable work of at least a decade of study, it could be said that Simona Stano describes, interprets, but above all makes us understand the fine mechanisms of that embrayage through which language and semiosis present their own nature as a naturality indistinct from nature itself, as a plane of language in which semiosis and natural reality, nature and meaning, are proposed and even touted as perfectly matching. This rhetoric—and Simona Stano demonstrates this perfectly in a series of explorations and probes that insist on her favourite territories, those of food, health and body discourses—is essentially 'adjectival', in the sense that, as we move from philosophical interdefinitions to the practices and especially the predicates of discourse, there is no longer any distinction between nature, whose adjectivisation is 'the natural', and naturality, whose adjectivisation also corresponds to 'the natural'. As Simona Stano perfectly demonstrates, 'natural' food is not so in the sense of nature, but in the sense of 'naturality', yet in the rhetoric of adjectivisation, one and the other converge to make us adhere to the idea that yes, between the fruit that grows in the Amazon rainforest, unseen by any human, and that which a cherished 'natural' restaurant serves us plated according to the gastronomic aesthetics of the moment, there is no distinction at all.

Therefore, they do make one smile, were one not irritated by their cunning but somewhat crude self-promotional intent, the 'against nature' arguments, as they fail to distinguish between nature and naturality and, ultimately, present rhetorics of unmasking that are just as rhetorical as those they aim to expose. Much more difficult, though perhaps farther from the spotlight, is the work of those who, like this fine book by Simona Stano, patiently weigh the discourses of nature and naturality, unravelling their innermost gears, without ever adopting paternalistic or moralistic attitudes, but instead doing proper semiotics, which is not, as many contemporary Solons believe, the art of unmasking, but that of dissecting. It is an anatomy of meaning, not its etiquette.

The lexicon that gravitates around nature, however, like Jupiter's many satellite mistresses, offers us not only the unattainable nature and its enunciated enunciation of naturality but also a third term, which in many languages is untranslatable, but which, fortunately, is present in both Italian (the language in which I think) and English (the language in which I write). This third word is 'naturalness', 'naturalezza'. It is no accident that, in Italian, 'naturalezza' rhymes with 'beauty', 'bellezza'. Naturalness is a naturality that has made it. It is not only an elegant embrayage, a successful enunciation, but also an attitude that has something childlike about it. It is no accident that in many languages it is rendered by the synonym 'spontaneity'. The Italians of the Renaissance also called it 'sprezzatura', a term with a thousand historical and semiotic implications, which unfortunately we cannot pursue here.

In naturalness, we do not deny the naturality of our meaning, but offer it to the world as if it were a fruit. In naturalness, we imitate nature sublimely. All sublime art is natural not in the sense of unattainable nature, nor in the sense of rhetorical naturality, but in the sense of a naturalness in which ancient myths hint there is like a shadow of the breath of creation. This distinction, too, is lost in adjectivisation, and we say that an exalted dancer moves his body with natural elegance; and yet philosophy and semiotics invite us to explicate the concepts behind the adjectives, to grasp that the natural elegance of an arabesque is not in the nature of the dancer's body (or at least not only in it) and is not even in its supposed naturality. How clumsy, in fact, a dancer would seem to us if he were to strive to appear 'natural'. 'Try to appear natural!' is indeed a phrase that could be added to the schizophrenic and schizogenic ones studied by the pragmatics of human communication. On the contrary, the dancer who makes us dream is natural in the sense of naturalness, and he makes us dream because, through the immense efforts that are not only his own as an individual dancer but also those of a whole history of dance that suddenly takes on body and nature in a calf, in a leap and in a pirouette, we catch not the elusive nature, nor the image of it that proposes to us a clumsy and often sly rhetoric of transparency, but a breath of creation and beauty, a gesture that, in artifice, whispers in our ears the nature of our best nature.

Massimo Leone
Department of Philosophy
and Education Sciences
University of Turin
Turin, Italy

Department of Chinese Language
and Literature
University of Shanghai
Shanghai, China

Director of the Centre for Religious
Studies
Bruno Kessler Foundation
Trento, Italy

in Europe that the general acceptance of anarchic [illegible] is not in the nature of the dancer's [illegible]

[illegible]

# Preface

This book derives from a particular interest that emerged from various studies I have conducted in recent years. While carrying out research on diverse topics related to food, corporeality and health, among others, something caught my attention: though focusing on different cases and contexts, I repeatedly encountered texts, discourses and practices revolving around a common idea, *nature*.

At first glance, this might not appear particularly surprising: as Noel Castree observed, 'nature is one of the most widely talked about ... things there is' (Castree 2005: xvii). In fact, who has not read at least one of the uncountable articles or online posts about 'natural' foods, 'natural' beauty products, 'natural' health treatments and so on and so forth? Who has not attended, physically or virtually, at least one debate on 'natural' resources and conservation practices? Who has not come across at least one of the manifold personifications of 'nature' promoted by literature, the arts, or the mass and new media? What is more, while these examples might seem particularly related to contemporary cultures and communication, 'nature has always been a major issue for societies worldwide' (*Ibidem*), as shown by a number of interesting reflections in the philosophical, sociological and also anthropological field, as well as in the artistic and aesthetic domain.

So, why a new book devoted to this issue? The reasons are multiple, but one is certainly more important than the others: today, more than ever, nature seems to have taken on a peculiar connotation, marking an evident gap between scholarly reflection and common sense. In fact, on the one hand, in recent decades academic research has seen important turning points in its approach to nature, with the appearance of new theories and models such as 'multinaturalism' and 'internaturality', based on the acknowledgment of the existence of different 'ontologies'—or 'ideologies', in a semiotic perspective—of nature, and the recognition of a variety of natures alongside the already recognised variety of cultures. On the other hand, a number of texts representing nature in Western societies seem not only to largely neglect such ideas but also to disregard the rich heritage of meanings and values that, over time, have been associated with the term 'nature', preferring a rather simplistic dichotomic view, which interprets it as a sort of common 'substrate' for all cultures, from which these are thought to have developed precisely 'by difference'—that is to say, by distinguishing themselves from it.

Two crucial questions therefore arise: how should we contextualise and understand the recurrent opposition between nature and culture in the discourses and practices that circulate throughout contemporary Western societies? And how can we—if we actually can—envisage alternative models and approaches capable of better accounting for the rich semantic scope of the term 'nature', as well as for the fundamental dynamics highlighted by recent academic research as regards its plurality and variability?

The pages that follow intend precisely to address these fundamental issues, combining the analysis of a wide range of relevant case studies with an in-depth review of existing literature in the considered fields of study. Thus, this book introduces a *critique of 'pure Nature'*, that is to say, a systematic study of the way nature is attributed meaning(s) and value(s), as well as particular forms and expressions, in some of today's most relevant discourses and practices, with particular reference to the evolution of such a concept over time and the attempt to trace a possible path towards an 'internatural turn'.

Turin, Italy Simona Stano

# Acknowledgments

This project has received funding from the European Union's Horizon 2020 research and innovation programme under the Marie Sklodowska–Curie grant agreement No 795025.

Its realisation was possible thanks to the kind support of the Department of Philosophy and Education Sciences of the University of Turin (UniTO), Italy, to which I am now particularly happy and honoured to be permanently affiliated, and the Steinhardt School of Culture, Education, and Human Development at New York University (NYU), United States, with which I had the pleasure and privilege to collaborate as a Visiting Research Scholar from January 2019 to December 2021. Furthermore, my research greatly benefited from the kind cooperation of a number of other academic institutions and research centres, which gave me the possibility to present its first results and further explore its multiple facets: the International Semiotics Institute (ISI), the International Association for Semiotic Studies (IASS-AIS), the Centre for Interdisciplinary Research on Communication (CIRCe), the Italian Association for Semiotic Studies (AISS), the CRT Foundation, the PhD in Sustainable Development and Climate Change and several universities and research centres around the world.

I am also extremely thankful to the colleagues and friends who have provided me with stimulating suggestions and enriching comments on various of the ideas presented in the following pages: Massimo Leone, Amy Bentley, Ugo Volli, Dario Martinelli, Antonio Santangelo, Jenny Ponzo, Anna Maria Lorusso, Franciscu Sedda, Maria Giulia Dondero, José Enrique Finol, Elena Casetta, Carolyn Dimitri, Krishnendu Ray, Mattia Thibault, Gabriele Marino, Bruno Surace, Vincenzo Idone Cassone, Federica Turco, Eleonora Chiais, Gabriele Vissio and several others.

Last, but certainly not least, I thank my family and friends—and especially Michela, Pietro, Cinzia, Marco, Laura, Maurizio, Alice, Simone, Maria (Pucci), Giovanna, Roberto, Ina, Gianni, Giovanna, Daniele, Federico, Angelo, Luisa, Alessandra, Matteo, Elisa, Laura, Nicola, Claudia, Stefano, Paola, Chloé, Isabelle, Lawrence, Matt, Kate, Marivi, Berta, Carla, Irene, Daniel, Umbe, Marco A.C., Sara, Ilaria, Luca, Jenny, Beatrice, Cristiano, Valeria, Paolo, Federica and Carlo, together with many others—for all their support, both in relation to this particular work and beyond it. Silvio and Dafne, you might be surprised not to find your names in this list, but that is just because I ought you my most special thanks, not only for your ever-present encouragement and patience (especially during the last months of intense writing!) but also and especially for how you make my every day so special and meaningful.

# Contents

# About the Author

**Simona Stano** (website: https://simonastano.it/en/) is an Associate Professor of Semiotics at the University of Turin (UniTO, Italy) and the Vice-Director of the Centre for Interdisciplinary Research on Communication (CIRCe). In 2018, she was awarded a Marie Curie Global Fellowship for a research project (COMFECTION, 2019–2021) on the semiotic analysis of food and digital communication.

She has collaborated with several universities and research centres in Italy and abroad, including New York University (USA, 2019–2021), the International Semiotics Institute (2015–2018), the University of Toronto (Canada, 2013), Kaunas University of Technology (Lithuania, 2015–2019), the University of Barcelona (Spain, 2015–2016) and Observatorio de la Alimentación (Spain, 2015–2016).

Prof. Stano deals mainly with food semiotics, body semiotics and communication studies, and has published several papers, edited volumes (including special issues of top semiotic journals such as *Semiotica*, *Lexia* and *Signata*), essays and monographs (*I sensi del cibo*, 2018; *Eating the Other. Translations of the Culinary Code*, 2015) on these topics.

# About the Author

# Chapter 1
# Introduction

> So, nature is not a physical place to which one can go, nor a treasure to fence in or bank, nor as essence to be saved or violated. Nature is not hidden and so does not need to be unveiled. Nature is not a text to be read in the codes of mathematics and biomedicine. It is not the "other" who offers origin, replenishment, and service. Neither mother, nurse, nor slave, nature is not matrix, resource, or tool for the reproduction of man.
>
> Nature is, however, a *topos*, a place, in the sense of a rhetorician's place or topic for consideration of common themes; nature is, strictly, a commonplace. We turn to this topic to order our discourse, to compose our memory. … Nature is also a *trópos*, a trope. It is figure, construction, artifact, movement, displacement. Nature cannot pre-exist its construction.
>
> Donna Haraway, *The Promises of Monsters* (1992).

**Abstract** This chapter provides an overview of the book's main topics and approach. Given the crucial role played by the concept of "nature" in present-day discourses, as some examples briefly discussed in the first paragraph demonstrate, its analysis will be carried out, recalling crucial references in relevant literature, in an interdisciplinary perspective (Chap. 2). Subsequently, the discursivisation of nature in different domains will be considered, ranging from the iconography of Mother Nature between the past and the present (Chap. 3) to the representation of catastrophic events in fictional and non-fictional discourses (Chap. 4), from "clean eating" and other popular contemporary food trends based on "natural mythologies" (Chap. 5) to the multiple meanings and values attributed to the naked body (especially in relation to the ambivalence between its supposed natural ascription and its multiple cultural characterisations) in different contexts (Chap. 6). Finally, Chap. 7 draws some conclusions on the analysed dynamics, also illustrating a possible path towards an "internatural" turn through a particularly relevant case study.



S. Stano, *Critique of Pure Nature*, Numanities - Arts and Humanities in Progress 26,
https://doi.org/10.1007/978-3-031-45075-4_1

## 1.1 In Praise of "Nature[1]": A Contemporary Myth

While always being central to our understanding of the world and of ourselves (as scholars like Claude Lévi-Strauss remarked on various occasions), the idea of "nature" now seems even to have become a sort of "categorical imperative", intended—with Kant's permission for such an enlarged and provocative use of this concept—as an "unconditional" and "absolute rule of conduct". Nowadays, in fact, it seems essential to give birth "naturally", to feed others (e.g. infants) and oneself "naturally", to use "natural" remedies for treating the body, and so on and so forth. A 2019 magazine article ironically referred to such a contemporary "myth of the natural" (Dangerfield 2019) by declaring: "If you want people to buy your product just label it *natural*. If you want to disapprove of something just label it *unnatural*" (*Ibidem*, my emphasis).

Indeed, while a series of "natural" substances (e.g. unrefined foods) are particularly acclaimed for their beneficial effects, "industrialised" and "modified" products are generally blamed for their dangerous consequences on both human health and the environment. Likewise, an increasing number of supporters of "natural" medical and body care treatments reproach the risks associated with chemical products and technological innovations, relating them to "unnaturality"; and so on and so forth. Although dating back to ancient times, such a "praise for Nature" has now reached an unprecedented diffusion, coming to encompass almost all spheres of everyday life, and finding expression in all sorts of communications. The myth[2] of a "pure Nature"—considered as the opposite of culture, namely as its absence and/or rejection—has thus been established, becoming a key component of Western contemporary cultures.[3]

However, as this book intends to show, this idea is highly problematic. Consider, for instance, the case of food, which is emblematic in this regard. Most of the products we eat, in fact, are not edible as they are found in nature, needing more or less complex processes of selection, processing and treatment to become suitable for human consumption. Leaving nature to take its course, moreover, would require us to avoid any process of preservation of food, as well as any system involved in its

[1] The word is capitalised to highlight the importance attributed to such a "principle" within the considered context, where it has become a sort of "secular deity" inspiring a number of trends, behaviours and practices. As a general rule, the capital initial will be used in the following pages to emphasise this type of conception (or similar ones), while the lower case will be preferred to refer to nature as a general idea.

[2] The term "myth" is here used to evoke the expression used by Elizabeth Dangerfield mentioned above, but also in relation to Roland Barthes' description of "mythology", which will be recalled and further described as a crucial idea for the project of this book (see *infra*, Sect. 1.2).

[3] In accordance with common standards, the expression "Western culture", sometimes also referred to as "Western civilisation" or "society", is used to denote a primary focus on the European context, as well as on those areas and cultures whose histories are strongly connected to Europe by immigration, colonisation or influence (e.g. North America). The plural form is preferred throughout this manuscript to highlight the diversity and the variety characterising such cultural contexts, which are fundamental for the analysis of the observed case studies.

packaging and transportation (which not only makes it arrive on our tables, but also keeps it—as well as those who ingest it—safe from external contamination).

Yet, today these cultural processes tend to be largely neglected, and sometimes even contrasted, in the name of an ideal of "pure Nature", which—as a series of relevant case studies related to different domains, from food itself to catastrophic events, etc. will reveal—is defined essentially according to a logic of denial, and a subsequent dysphorisation, of any "artificial", "altered", "manipulated"—in short "cultural"—aspect. In such a view, as it will be shown, Nature is mainly what things *are not*, or *not yet*, in a simplistic and stereotyped—as well as stereotyping—view.

Thanks to an original theoretical and methodological approach, the pages that follow provocatively challenge such a dichotomic view, questioning the concept of "pure Nature" supposed by most contemporary discourses in a critical perspective.

## 1.2 For a Critique of Pure Nature: Main Purposes and Structure of the Book

Developing a "critique of pure Nature", as this books aims to do, means first of all systematically enquiring into the conditions and consequences of such a concept, in order to understand its *possibilities*, or conditions of existence, as well as its *validity*, and especially its *limits*.

In order to do so, Chap. 2 introduces an in-depth analysis of the term "nature", opening the way to a systematic reflection on decisive moments in the evolution of such a concept over time, ranging from classical philosophy (and especially Aristotle's description of *physis*, which has had a great influence on Western thought) to the crucial changes occurred through the Middle Ages, the Renaissance, Romanticism and the modern era, and finally considering more recent insights from leading scholars in philosophy, sociology and cultural anthropology, as well as in the semiotic field. While not seeking to provide a comprehensive overview—which would be an impossible goal, especially in such a reduced space—, this section helps illustrate the problematic aspects related to the definition of nature, highlighting the ambivalences and limits of this idea and, more specifically, of its use in contemporary Western cultures.

Hence, the book explores the contemporary *mythologies*[4]—or, to adopt a more common terminology, the *ideologies*[5]—of Nature, that is to say, the particular connotative operations through which certain values associated with such a concept are attested and promoted, while others are narcotised. This means recognising nature as a "discursive value", in the aim to "*de*-naturalise" it, namely to analyse how it is figurativised and attributed meaning, with particular reference to Western cultures. To this purpose, a series of relevant case studies are analysed through a semiocultural approach, describing how Nature is thematised and given specific values in various texts and discourses circulating within contemporary societies (also in relation to traditional and historical representations and valorisations).

More specifically, Chap. 3 focuses on the collective imaginary of Mother Nature, from classical iconography to contemporary popular culture (e.g. music, cinema, television and advertising). This also allows highlighting crucial aspects for the development of the following chapters, which build precisely on some of the disclosed elements to enhance the reflection on the main aspects of the current discursivisation of Nature in various domains.

Chapter 4 deals with the analysis of the collective imaginaries of catastrophic events, ranging from fictional discourses, with a particular focus on movies, to non-fictional representations. The discussion of relevant case studies points out the interconnection between "natural" disasters or calamities, on the one hand, and "cultural" catastrophes, on the other hand: while the former seem to be at least partially related to ecological or social dislocations caused by (sometimes unconscious, sometimes irresponsible) human actions, the latter can be conceived as natural entropic degenerations of complex organisations. However, as most narratives of the so-called "Anthropocene" and various environmental catastrophes, as well as a number of discourses related to the recent Coronavirus pandemic, clearly show, a dichotomic view of the relation between nature *vs.* culture seems far from be abandoned.

Chapter 5 deals with the food universe, within which the concepts of nature and culture have always played a crucial role. After an initial theoretical framing recalling Claude Lévi-Strauss' reflections on cooking and the evolution of cooking systems and their significance over time, some of today's most popular food trends (from "clean eating" to aphrodisiac substances, from genetic manipulations to the so-called organic paradigm) are analysed through a semiocultural approach in order to investigate the special praise for "uncooking" processes that seems to characterise

---

[4] The term "mythology" is used in the sense described by Barthes (1957), that is to say, as a "second-order semiological system" or "metalanguage" that exalts certain values and narcotises others, naturalising specific visions of the world. For a further discussion of this idea, especially in relation to contemporary cultures, see in particular Stano (2022, 2023).

[5] The term is used as an equivalent of "mythology" (or "myth") based on its semiotic understanding as a connotative process—and more specifically, as "the final connotation of the totality of the connotations of the sign or context of signs" (Eco 1968: 96)—which "conceals" the traces of its original enunciation, obstructing the meta-semiotic function of discourse (cf. Eco 1976). For a further discussion on ideology as related to meaning-making processes, and in particular on the analogy between Barthes' description of myth and Eco's reflections on ideology, see in particular Stano and Leone (2023).

contemporary foodspheres, revealing a series of inconsistencies and problematic aspects related to it.

Chapter 6 provides insights on the ambivalence of the naked body between its supposed natural ascription and its multiple cultural characterisations. In fact, as a popular expression says, we are all born naked, "just as nature intended". However, various meanings have been attributed on nudity over time, as the chapter shows through a diachronic overview. Relevant examples of the use of nakedness as a powerful tool in political action, artistic performance, and a series of other significant practices aimed at (re)writing and (re)semantising the corporeal dimension for different purposes are also analysed, pointing out the effects of meaning arising from them, especially as related to the relation between nature and culture (and the way these poles are discursivised).

Finally, Chap. 7 draws the conclusions of the research, reconsidering the ambiguities and limits of the conception and description of Nature in Western thought in the attempt to move forward towards an "internatural turn" and a better comprehension of natures—now in a plural and, definitely, non-capitalised form—, also based on a particularly interesting case study observed during the final months of the research.

Thus, the critique of pure Nature presented in the following pages is intended to show that nature is not—to recall Donna Haraway's reflections reported in the epigraph above—"the 'other' who offers origin, replenishment, and service. Neither mother, nurse, nor slave, nature is not matrix, resource, or tool for the reproduction of man" (Haraway 1992: 296). Rather, it is "a *topos*, a place, in the sense of a rhetorician's place or topic for consideration of common themes; nature is, strictly, a *commonplace*" (*Ibidem*, my emphasis), as well as a "*trópos*, a trope. It is figure, construction, artifact, movement, displacement. Nature cannot pre-exist its construction" (*Ibidem*). As such, as this book illustrates, it is as much part of culture as culture is of it.

## References


Barthes, Roland. 1957. *Mythologies*. Paris: Editions de Seuil. English Trans. Lavers, Annette. 1972. *Mythologies*. New York, NY: The Noonday Press—Farrar, Straus and Giroux.

Dangerfield, Elizabeth. 2019. The myth of the natural. *Areo*, Aug 08. https://areomagazine.com/2019/08/05/the-myth-of-the-natural/. Accessed 21 Dec 2022.

Eco, Umberto. 1968. *La struttura assente*. Milan: Bompiani.

Eco, Umberto. 1976. *A Theory of Semiotics*. Bloomington, IN: Indiana University Press.

Haraway, Donna. 1992. The promises of monsters: A regenerative politics for inappropriate/d others. In *Cultural Studies*, ed. Lawrence Grossberg, Cary Nelson, and Paula A. Treichler, 295–337. New York-London: Routledge.

Stano, Simona. 2022. Il mito tra verità e post-verità/Myth between truth and post-truth. *RIFL* 2022 (SFL2021): 204–213.

Stano, Simona. 2023. Nuovi media, nuovi miti? In *Nuovi media, nuovi miti*, ed. Jenny Ponzo and Simona Stano, 21–35. Rome: Aracne.

Stano, Simona, and Massimo Leone (eds.). 2023. *Ideologia—Ideology*. Special issue of *Lexia* 41–42. Rome: Aracne.

# Chapter 2
# (Re)Defining Nature: From Praise to Critique

**Abstract** This chapter provides an analysis of the concept of "nature", opening the way to a systematic reflection on decisive moments in its evolution over time, ranging from classical philosophy to the crucial changes occurred through the Middle Ages, the Renaissance, Romanticism and the modern era, and finally considering more recent insights from leading scholars in philosophy, sociology and cultural anthropology, as well as in the semiotic field. While not seeking to provide a comprehensive overview (which would be an impossible goal, especially in such a reduced space), the following pages help illustrate the problematic aspects related to crucial changes in the definition of nature, highlighting the ambivalences and limits of such a concept and especially of its use in contemporary Western cultures.

## 2.1 The Term "Nature" and Its (Problematic) Polysemy

As John Habgood wrote, "an immediate problem with the word 'nature' is that it has multiple and overlapping meanings" (Habgood 2002: 1). The Merriam-Webster Dictionary, for instance, includes more than ten different points in its description of the term:

> 1. the external world in its entirety; 2. natural scenery ("enjoyed the beauties of *nature*"); 3a. DISPOSITION, TEMPERAMENT ("it was his *nature* to look after others"—F. A. Swinnerton; "her romantic *nature*"); b. the inherent character or basic constitution of a person or thing: ESSENCE ("the *nature* of the controversy"); 4a. humankind's original or natural condition: b. a simplified mode of life resembling this condition ("escape from civilization and get back to *nature*"); 5. a kind or class usually distinguished by fundamental or essential characteristics ("documents of a confidential *nature*"; "acts of a ceremonial *nature*"); 6. the physical constitution or drives of an organism (especially, an excretory organ or function, used in phrases like "*the call of nature*"); 7. the genetically controlled qualities of an organism ("nature … modified by nurture"—E. G. Conklin); 8a. a creative and controlling force in the universe; b. an inner force (such as instinct, appetite, desire) or the sum of such forces in an individual; 9. a spontaneous attitude (as of generosity) (Merriam-Webster 2022: s.v. "nature").



S. Stano, *Critique of Pure Nature*, Numanities - Arts and Humanities in Progress 26,
https://doi.org/10.1007/978-3-031-45075-4_2

The first element highlighted in this definition evidently recalls the etymology of the term "nature", which derives from the Latin *nātūra*, an equivalent[1] of the Greek *φύσις*, *physis* (Dunshirn 2019), in turn derived from the verb *φυεσθαι/φυναι*, *phyesthai/phynai*, "to (cause to) grow", "to develop" (see Frisk 2006: 1052; Caspers 2010: 1068)[2] and related to the root *γεν-*, "to come into being", "to become" (see Heidegger 1976 [1939]: 239).[3] In this sense, nature is mainly conceived as a generating principle (as confirmed by its description in terms of a "creative and controlling force" in point 8a), as well as the totality of being (as remarked by the expression "in its entirety" used in the definition), recalling the ambivalence between the idea of a *natura naturans*, i.e. a "creative" power, and a *natura naturata*, i.e. the "created" world, often used during the Middle Ages and the Renaissance and then recalled by philosopher Spinoza (who associated the former with God and distinguished the latter into universal and particular, see Spinoza 1994: 57–58, cf. Jaspers 1974: 14, 95).[4]

The adjective "external"[5] (point 1) interestingly suggests a separation between the natural world and the potential readers (i.e. humans) of this description, which is also reinforced by the reference to natural sceneries (suggesting a look directed at them) in point 2. Hence, while a "cosmological" sense[6]—as described by Soper (1995: 21)—is partially recalled, a separation between nature and humanity is also present—and particularly remarked throughout the definition, and especially in point 4b, which explicitly contrasts nature with "civilization", and point 7, which counterposes it to "nurture" (recalling human intervention) in the sentence provided as an example. Such an opposition is further emphasised by the definitions provided by other dictionaries, as the case of the Cambridge Dictionary clearly exemplifies: "all the animals, plants, rocks, etc. in the world and all the features, forces, and processes that happen or exist *independently of people*, such as the weather, the sea, mountains, the production of young animals or plants, and growth" (Cambridge Dictionary 2022: s.v. "nature", my emphasis).

In fact, as Soper highlighted,

> In its commonest and most fundamental sense, the term 'nature' refers to everything which is not human and distinguished from the work of humanity. Thus 'nature' is opposed to culture,

---

[1] However, as Dunshirn remarked, "with the Latin *natura*, which for its part goes back to the verb *nasci* ("to be born"), one transfers the basic word *physis* into a different sphere of association. In this way, the emerging growth (of plants, for instance) is transferred into the realm of being born" (2019: 1)—as Heidegger (1976 [1939]) also highlighted, see *infra*. For a more detailed discussion on the historical and semantic peculiarities of the Latin term, see in particular Pellicer 1966.

[2] With the suffix indicating the realisation of an abstract concept (Benveniste 1948).

[3] For the analysis of the various meanings associated with the term in Greek culture, see *infra*, Sect. 2.2.

[4] For further details, see *infra*, Sect. 2.4.

[5] In this respect, it is interesting to recall Noel Castree's description of the principal meanings of the word "nature" in contemporary Anglophone contexts, which refers to "the non-human world of living and inanimate phenomena" (Castree 2014: 10) precisely by means of the expression "external nature".

[6] Corresponding to the idea of "universal nature" described by Castree (2014).

to history, to convention, to what is artificially worked or produced, in short, to everything which is defining of the order of humanity (Soper 1995: 15).

As the next chapters will illustrate in detail, despite various problematic aspects, this is indeed the dominant view in contemporary Western collective imaginaries. Nonetheless, as Ducarme and Couvet (2020) remarked, such a view originated rather recently, especially as a result of the Scientific Revolution and the emergence of modern science and post-Romantic philosophy in early modern Europe.

Before moving to a more detailed description of the historical evolution of the concept of nature, however, it is worthwhile focusing on another fundamental aspect that emerges from the definition reported above, namely the conception of nature as the "essence"—i.e. an inner quality, a character, or a disposition—of things (as mentioned in points 3b, 5 and 6), and also of people (as clarified in these same points, and further emphasised by the descriptions provided in points 3a, 8b and 9). As Martin Heidegger observed, in fact,

> We likewise speak of the "nature" of spirit, the "nature" of history, and the "nature" of man. By the last phrase we mean not just man's body or even the species "man," but man's whole "Being." Therefore generally when we speak of the "nature of things," we mean *what* things are in their "possibility" and *how* they are, regardless of whether and to what degree they "actually" are (Heidegger 1976 [1939]; Engl. Trans. 1976: 222).

In this sense, man and nature are no longer opposed, "for we, too, it is said, are possessed of a 'nature', and may behave in more or less 'natural' ways" (Soper 1995: 25)—with the interpretation of such "natural ways" depending on very divergent and often antithetical moral postures and political ideologies (see *Ibid.*: 25–33). Heidegger himself insisted on this aspect, recalling the diversity between Christian thought—according to which human nature "brings about, through the passions, the total destruction of man" (Heidegger 1976 [1939]; Engl. Trans. 1976: 222), and should therefore "be held down" (*Ibidem*)—and Nietzsche's philosophy—which rather emphasises the "naturality" of man's unleashing of drives and passions, by virtue of which the *homo natura* is able to "sense" the world, and hence to master it.

As this brief overview clearly shows, a considerable "polysemic halo" surrounds the word *nature*, making it difficult to isolate an unequivocal meaning (cf. Arias-Maldonado 2015: 17; Larrère and Larrère 2015) and highlighting the risk for it to become "another meaningless *panchreston*"[7] (Ducarme and Couvet 2020: 1; cf. Vogel 2015: 9). In fact, the varied and sometimes even conflicting denotations recalled above represent a heritage of its history, which has made it go through various changes (cf. Lenoble 1969), including or excluding mankind and "culture", and reflecting different views and ideologies. An increasing number of linguists, philosophers and historians have thus insisted on the complexity and ambiguity of the term, highlighting the "constructed" character of what we call "nature":

[7] Drawing on Daniel Simberloff's reflection on the metaphor of the "balance of nature" (Simberloff 2014), Ducarme and Couvet (2020) extended the idea of a too broadly inclusive and often oversimplified expression (i.e. a panchreston, from the Latin *panchrēstos*, "good for everything, universal") to "nature" itself.

> There is perhaps something inherently mistaken in the attempt to define what nature is, independently of how it is thought about, talked about and culturally represented. There can be no adequate attempt … to explore 'what nature is' that is not centrally concerned with what it has been *said* to be, however much we might want to challenge that discourse in the light of our theoretical rulings (Soper 1995: 21).

As reported in Chap. 1, this is precisely the objective of this book. Before doing so, however, it is advisable to recall at least some of the most significant moments in the evolution of the concept of nature, ranging from classical philosophy to the crucial changes occurred through the Middle Ages, the Renaissance, Romanticism and the modern era, and finally considering more closely some recent insights from leading scholars in philosophy, sociology, cultural anthropology and semiotics.

## 2.2 Nature in Classical Thinking: From the Greek φύσις to the Latin *Nātūra*

As mentioned above, the origin of the word that later got translated into "nature" is generally traced back to the Greek term φύσις, *physis*. The oldest known mention of this term can be found in Homer's *Odyssey* (c. 8th century BCE), where it is used to describe the specific "character" of plants, designating their "appearance" (as maintained by Ducarme and Couvet 2020: 2)—i.e. the "created material world" (see *supra*)—, but also the "creative" forces underlying them (as highlighted, among others, by Arias-Maldonado 2015: 21). In fact, as Gerard Naddaf remarked, "in Homer, *phusis* designates the whole process of growth of a thing from its birth to its maturity" (Naddaf 2005: 16).

Such a conception seems to find a correspondence in the particular interest of Presocratic philosophy (6th and 5th century BCE) in cosmogony,[8] intended as the history (i.e. a *dynamic* process) of the universe, rather than its simple (*static*) description: "an explanation of its origin (*phusis* as absolute *archē*), of the stages of its evolution (*phusis* as process of *growth*) and finally of its result, that is, the *kosmos* as we know it (*phusis* as the result)" (*Ibid*.: 20). Nature therefore emerges as an *active subject*, "which does things, has goals, and pursues specific values"[9] (Marrone 2015: 2, my translation).

What is more, according to Heraclitus' well-known aphorism *physis kryptesthai philei* (fragment 123), such a subject "loves to"—or, better, "is accustomed to"[10]— "hide". While, as Hadot (2004) remarked, the meaning of these seemingly simple three words is not clear yet (despite the considerable efforts by various scholars to

[8] According to Naddaf (2005: 35), what differentiates the use of the term by Presocratics from its Homeric ancestor is precisely the reference to the gods.

[9] "Qui fait des choses, a des objectifs et poursuit des valeurs spécifiques".

[10] In fact, while the expression "to love" is frequently used to translate the saying, the Greek verb *philein* was rather used in the sense of "being accustomed to" by Heraclitus, as well as by Herodotus and other ancient authors, denoting "not a feeling but a natural or habitual tendency, or a process that occurs necessarily or frequently" (Hadot 2004; Engl. Trans. 2006: 7).

decipher it), and will probably never be, it has been largely interpreted as a reference to the "secrets" of Nature, resulting in two main tendences[11]: a *Promethean* attitude, aimed at discovering such secrets by "unveiling" Nature, in order to dominate it (especially when its effects go against men's interests); and an *Orphic* attitude, rather based on a sound respect for the mysteries and inaccessibility of Nature, which therefore rejects any form of "violence" or forced revelation in the aim to become intimate with it, penetrating its truths through aesthetic perception, artistic expression and philosophical discourse.

The agency of Nature is also particularly emphasised in Aristotle's reflections on φύσις, which have been fundamental in shaping most of the meanings that are still present in common language nowadays (see in particular Lammer 2015; Owens 1968; Ducarme and Couvet 2020). In the second book of *Physics*, the Greek philosopher defined nature as the "essence" of things, referring to what they are made of and, at the same time, what entails their destiny.[12] In another passage, Aristotle recognised the partiality of this definition, subsequently coming to include as many as four rather divergent descriptions of nature in *Metaphysics* (Δ4, 1014b): "the generation of what grows (as a *process*), the primordial element from which things grow (as a *principle*), the principle of movement (a *spontaneous cause*), and the matter from which things are made (*substance*)" (Ducarme and Couvet 2020: 2, my emphasis). While an analogy clearly emerges between these descriptions and the contemporary "cosmological" definition of nature (see *supra*), as well as the idea of a particular "essence", "disposition" or "physical configuration" of the world, no element seems to explicitly recall the opposition between nature and mankind.

Conversely, in the second book of *Physics* (II, 1–2), a reference is actually made to the contrast between φύσις, on the one hand, and τέχνη (*téchnē*)—that is to say, art and craftwork, or in a broader sense, "culture"—, on the other hand. In Aristotle's view, while natural entities have a source of dynamism and transformation in themselves, artefacts (i.e. the products of τέχνη) have it due to the activity of a human maker. Hence, they are considered simple inert tools, whose ontological status is linked to their use and the utility they have for man, who is for them the "principle of movement", their efficient cause (see also Licata 2019: 15). However, as various scholars have noted, both of these movements produce entities by imposing a form on matter according to a finalised plan. What is more, in such a view, "mankind remains [in any case] a part of nature, though able of making artifices" (Ducarme and Couvet 2020: 2). If an opposition exists in the Aristotelian model, then, it is rather between nature and "chaos", as "Nature is everywhere a cause of order" (*Physics*, 252al2)—which interestingly subverts the opposition between "civilization" and nature highlighted by contemporary definitions (see *supra*).[13]

[11] As Hadot remarked, such attitudes can follow one another, coexist, and even mix with each other, while remaining opposite. For a detailed description of some of their most relevant expressions over time, see in particular Parts V and VI of *The Veil of Isis* (Hadot 2004).

[12] As Casetta (2020) pointed out, it is in this sense, for example, that we speak of "human nature".

[13] As Ducarme and Couvet highlighted, in fact, "civilized men are more 'natural' in this point of view, as they live under laws, than 'barbarian' peoples, submitted to disorder and then oblivious of

The contrast between nature and culture arose more evidently in the Roman context. Infrequently used until the 2nd century BCE, mostly as related to the idea of "birth" and "origin" (i.e. in its etymological sense[14]), the Latin term *natura* initially inherited its full philosophical, Greek-influenced meaning in the following century (Pellicer 1966), primarily through the work of Lucretius, who expounded the Epicurean view that all things, from the microscopic world of atoms to cosmic phenomena, occur by natural causes. In fact, the poet's *De rerum natura* aimed precisely at "explain[ing] by what forces nature steers the courses of the sun and the journeyings of the moon, so that we shall not suppose that they run their yearly races between heaven and earth of their own free will with the amiable intention of promoting the growth of crops and animals, or that they are rolled round in furtherance of some divine plan" (*De rerum natura* 5.76–81). It was then Cicero who significantly innovated such a conception, emphasising precisely the opposition between nature, intended as "an initial state devoid of human influence" (Ducarme and Couvet 2020: 2), and culture, rather defined by the "appropriation" of nature by human societies. As Ducarme and Couvet (*Ibidem*) noted, this was also reflected in the way the Romans perceived space: contrary to the traditional euphoric Greek view of the *polis*, Roman cities (i.e. urbanised, or "culturalised", spaces) were generally conceived as places of filth and sin, as opposed to the "good life" of the bucolic (i.e. "natural") countryside.[15] Such a view was then further enhanced by the advent of Christianism, which projected a "holy" aura on natural, "wild" spaces (conceived as the place of encounter with God, symbolised by the Garden of Eden),[16] as opposed to the "evil" of cities (e.g. the corruption of Babylon).

This contributed to increase the ambiguity and variability of the idea of nature, prefiguring a collective imaginary that has further developed and spread over time—as the following chapters will illustrate. However, as the authors of "What does 'nature' mean?" remarked, "nature and culture were still seen as dynamic processes rather than fixed states" (*Ibid.*: 3).

their human nature (a man living like a beast is as unnatural as a beast living like a man) (Lenoble, 1969). This is why 'nature' is not a synonym of wild, wildness or wilderness" (Ducarme and Couvet 2020: 2).

[14] See *supra*.

[15] An idea that, as it will be discussed in detail in Chap. 4, is far from having disappeared.

[16] Indeed, wilderness has both a positive and a negative connotation in Biblical texts: while the "Terrestrial Paradise" is said to have been created by God "to grow every tree that is pleasant to the sight, and good for food" (*Genesis* 2, 9), with no requirement for man other than respecting and protecting it, after the Fall it becomes a "disorderly wilderness", requiring to "be redeemed thorough cultivation and domestication" (Merchant 2003: 19). For a more detailed analysis of the evolution—and resemantisation—of the concept of wilderness, see in particular Oehlschläger (1991) and Ljungberg (2001).

## 2.3 Nature Between the Middle Ages and the Renaissance

As mentioned above, on the one hand, Christianism fostered a marked euphoricisation of the natural environment, as opposed to its transformation and excessive exploitation by man. On the other hand, however, it also resulted in a process of "desacralisation" of nature:

> In the Old Testament, a distinction between the divine and the natural is to be found, signalling that the Christian God is not immanent but transcendent: He has created nature, but He is not nature (Passmore 1974) (Arias-Maldonado 2015: 22).

Unlike the divine figures of Greek and Roman cosmogonies, the Christian God was not conceived as part of Nature, but rather seen as the creator of the natural world, which therefore represented a "simple (though magnificent) tool in [His] hands" (Ducarme and Couvet 2020: 3). Such an idea of a domination of the divine will over nature, which was therefore seen as impenetrable and unpredictable, became dominant during the Middle Ages: men could do nothing but pray God to spare them from adverse climate conditions and other devastating events, which were interpreted as a punishment for their sins. This fatalistic view was paired, on a philosophical level, with a conception of nature as essentially material, and therefore decidedly less important than the spiritual dimension.

This conception radically changed during the Renaissance, based on the belief that man could actually know and shape nature (cf. Callicott and Ames 1989). Nicholas of Cusa (1440), for instance, described nature as the *explicatio* ("unfolding") of God, meaning that the divine is present in nature as the "soul" of the world, which guides it from within and enlivens it, mediating between (and unifying) the maximum of unity, which is God himself, and the maximum of individuality, represented by matter. In his view, in other words, every being is simultaneously arranged by such a soul of the world and subjected to mechanical laws. Accordingly, for Ficino (1489), the soul of the world penetrates nature, conferring "intelligence" and "purpose" on it, and shaping it according to a preordained design, so that every living being manifests an internal planning that guides its development and allows its understanding and control. In his *De rerum natura iuxta propria principia* (1971 [1565–86]), Bernardino Telesio also insisted on the fact that since nature is a creation of God, who established its laws, and God cannot contradict himself, the laws of nature are necessarily stable and predictable—which assures man, once he has understood them, the possibility of controlling the natural order.

The dynamic idea of Nature as a creative process thus progressively left the way to a more static conception, which interpreted it as an initial state (cf. Simberloff 2014), whose possibility of transformation was attributed to man. As a result, a marked process of resemantisation of "culture"—intended as the human intervention on nature—occurred, also coming to encompass the sphere of *magic*.

More specifically, between the High Medieval Period and the Renaissance, this passed from being depicted as a "pact with demons" (as it was, for instance, in Augustine of Hippo's *De civitate Dei*, 21.6) to being celebrated as a "natural affinity", as Marsilio Ficino commented in his famous work *Sopra lo amore ovvero Convito di Platone*:

> The work of magic is the attraction of one thing by another because of a certain affinity of nature. But the parts of this world, like the parts of a single animal, all deriving from a single author, are joined to each other by the communion of a single nature. Just as the brain, lungs, heart, liver and the rest of the parts draw something from each other and sympathize with any of them when it suffers, so the parts of this great animal, that is all of the bodies of the world, similarly joined together, borrow and lend natures to and from each other. From this common relationship is born a common love; from love a common attraction. And this is the true magic (Ficino 1484; Engl. Trans. 1985: 127).[17]

Magic thus arose as "the natural philosophy par excellence" (Hadot 2004: 112), which also made it approach the realm of science, as well as that of natural phenomena *tout court*, juxtaposing transformations of various kinds:

> Thus, the magnet attracts iron, amber attracts straw, and sulfur fire. The sun makes many a flower and leaf turn toward it; the moon has the custom of attracting water, Mars the winds, and various herbs also attract various kinds of animals to themselves. Even in human affairs, each one undergoes the attraction of his own pleasure. The works of magic are therefore works of nature ... (Ficino 1484; Engl. Trans. 1985: 127).

While remaining part of nature, therefore, man was particularly praised for his ability to dominate it (see also Agrippa von Nettesheim 1533), in a vision that was destined to grow, opening the way to new understandings and politics of nature.

## 2.4 Nature Between the Scientific Revolution and Romanticism

As a result of the changes emerged during the Middle Ages and the Renaissance, the opposition between nature and "culture"—intended as the transformation of the former by man—further increased, leading to a dualistic and mechanistic vision of nature, which largely spread throughout Europe through the ideas and works of philosophers and scientists like Francis Bacon, René Descartes and Isaac Newton. Conceiving nature as inert matter regulated by definite physical laws,[18] detached from any divine principle or character, such thinkers radicalised the dichotomy "man" *versus* "nature", explicitly connecting it with a particular hierarchy: the former stood above the latter, which was thought as composed of objects metaphysically separated from humanity (cf. Pepper 1996).

As a result, a marked Promethean attitude towards nature imposed itself, based on the idea of a machine (an *object*) controlled by human beings (*subjects*), which

[17] Such an organicist view can be found also in the work of Tommaso Campanella (1620), who insisted precisely on magic in the comprehension of the links between the world constituents.

[18] Descartes (1664, Chap. 7), for instance, described it as "matter itself", explicitly rejecting any association with "some deity or other sort of imaginary power" (*Ibidem*), while Bacon insisted precisely on the fact that "nothing exists in nature, except individual bodies, exhibiting clear individual effect according to particular laws" (Bacon 1620, II, Aph. 2, 3–4).

were rather seen as "the crown of creation, the source of all value, the measure of all things" (Grierson 2009: 199). The result, as Merchant (1980) provocatively claimed, was the "death of nature", intended as both the collapse of the very concept of nature, and the subjugation of the natural world to scientific and economic interests.[19] In fact, several scholars have insisted on the role played by such a vision in the process of "unveiling" (to recall the terminology used by Hadot 2004), appropriation and exploitation of nature that has marked the modern and contemporary era (see in particular the works by Latour and Descola, whose main ideas are considered in further detail in the following paragraph), also forging a collective imaginary based on a sharp separation and opposition between the natural world and the anthropological and cultural dimension.

However, it should be noted that, while minor, a deistic and more markedly Orphic vision of nature also persisted alongside the prevailing Promethean one. In various of his works (and particularly in *Ethics*, the *Short Treatise on God, Man, and His Well-Being* and *Metaphysical Thoughts*[20]), for instance, philosopher Spinoza recalled the ambivalence between a *Natura naturans*, or "naturing Nature", conceived as an infinite generative continuum through which all finite beings come to existence, and a *Natura naturata*, or "natured Nature", intended as the actualisation of such a potential continuum in specific "modalities" (i.e. ways of being, thinking and acting), emphasising the connection between the natural world and the divine sphere. In such a perspective, "Nature is *one*, from the *conjunctive* view of substance and attributes; nature is *many*, from the *disjunctive* view of modalities. One, but all-varying. Multiple, yet all-connected" (Kodalak 2020: 125). Transcendent, though also immanent. Subject, but also object.[21] This entailed a more respectful attitude towards the mysteries of Nature, rejecting any attempt to forcefully reveal them. More generally, as Lotman (1967) remarked, during the Enlightenment period the opposition "natural" *versus* "unnatural" was elected as a general foundation for the organisation of culture, recalling the contrast between the reality of the world of "things", on the one hand, and the deceitful character of social and cultural signs and processes, on the other hand.

Thus, from the mid-eighteenth century, the attempt to unveil nature started vanishing, leaving the way to a sense of wonder, and at the same time of fear, towards Nature—which found expression, for instance, in Goethe's verses "Mysterious even in open day/Nature retains her veil, despite our clamours/That which she doth not willingly display/Cannot be wrenched from her with levers, screws, and

[19] Merchant also highlighted the link between nature and the female sphere, thus extending these processes to the role and representation of women in society.

[20] *Ethica, ordine geometrico demonstrate* (1661–1675), *Korte Verhandeling van God, de Mensch en deszelvs Welstand* (1661–1662), and *Cogitata metaphysica* (1663); all available in English in Curley (1985).

[21] Yet, as Descola (2011) pointed out, this model failed to include "the intermediary states, the compromises, the different forms of conciliation" between "the rigorous naturalis[m] of nature naturing and the staunch culturalis[m] of nature natured. Many geographers, sociologists, anthropologists, and philosophers have strived to find a dialectical move that would allow them to sidestep the confrontation between these two dogmatisms" (Descola 2011; Engl. Trans. 2013: 28–29).

hammers" (*Faust*, Part I, Scene I). Evidently breaking with Mechanism, such a vision "emphasize[d] diversity over unity, heterogeneity over homogeneity, the whole over the parts" (Arias-Maldonado 2015: 24). In fact, in 1753 philosopher Denis Diderot published his *Pensées sur l'interprétation de la nature*, where he insisted precisely on the impossibility for man to grasp the diversity and complexity of Nature, paving the way for its Romantic conception as a dynamic and incredibly rich whole, which cannot be reduced to the sum of its parts, and, therefore, should not be rationalised, nor instrumentalised.

Though with evident differences, such a conception found some correspondence in Charles Darwin's evolutionary theory, especially as related to the complexity and dynamism of the natural order. In Darwin's view, nature is "a web of complex relations and ecological interdependences: a grand scheme of cooperative integration… an innovative, reproductive force" (Arias-Maldonado 2015: 26). Where there is nature, there is evolution. However, it must be noted that evolution does not represent the domain of necessity, but rather that of potentiality, or of choice, and therefore of freedom, incalculability, and imponderability. In other words, as Massimo Leone highlighted, "it is in the very nature of nature, evolutionarily understood, that it must be imagined freely, that is to say, based on modalities that are not logical, but rather semiotic"[22] (Leone 2012: 201, my translation).

## 2.5 Nature in Contemporary Cultures: Main Concepts and Still Open Issues

While certainly not exhausting the multiple and diverse definitions of nature that have been developed over time, the previous paragraphs recalled the most important changes in its understanding up until the 19th century, pointing out the richness, and at the same time the extreme ambiguity, of this concept. In fact, already in 1874, John Stuart Mill lamented how "unfortunate" it was that terms such as "nature", "natural" and other expressions derived from them,

> Which play so great a part in moral and metaphysical speculation, … ha[d] acquired many meanings different from the primary one, yet sufficiently allied to it to admit of confusion … and entangled in so many foreign associations, mostly of a very powerful and tenacious character, that they ha[d] come to excite, and to be the symbols of, feelings which their original meaning will by no means justify, and which have made them one of the most copious sources of false taste, false philosophy, false morality, and even bad law (Mill 1904 [1874]).

As Ducarme and Couvet (2020) illustrated, similar warnings also appeared in the *Encyclopédie* (e.g. "this rather vague word, often used but hardly defined, that philosophers tend to use too much", D'Alembert et al. 1765: s.v. "nature"), as well as in other reflections by naturalists (such as Buffon 1770) and philosophers (such as

[22] "È nel carattere stesso della natura, evoluzionisticamente intesa, di dover essere immaginata liberamente, ossia secondo modalità non logiche ma semiotiche".

Merleau-Ponty 1995). Accordingly, Carolyn Merchant described nature as the result of the various ethics (and especially the "egocentric, homocentric, and ecocentric ethics", Merchant 2005 [1992]: 61) developed in Western culture since the 17th century, interestingly highlighting the role of modern "political, religious, and ethical trends" (*Ibidem*) in shaping our understanding of and relation to nature.

Despite such a variety, as noted above, a particular aspect seems to have progressively imposed itself in such a plethora of meanings and conceptions: a *differential view* of nature as the opposite of culture, namely as what chronologically and ontologically "precedes" it. As Jean-Marc Besse (2004) remarked:

> Until now ..., whatever form their relationship took, nature was thought of as first, chronologically and ontologically, in relation to culture. Culture came after nature, which was, so to speak, its 'setting'[23] (my translation).

Whether seen as a domain based on regularity and stability, or rather conceived as the realm of spontaneity and instinct, Nature has been increasingly defined as "what Culture is not" (or "not yet"), always disregarding—as Maurice Merleau-Ponty interestingly remarked in his well-known lectures on the concept of Nature (1956–1960)[24]—that nature "is not really set out in front of us. It is our soil [*sol*]—not what is in front of us, facing us, but rather, that which carries us" (Merleau-Ponty 1995; Engl. Trans. 2003: 4). Rather than pointing at the opposition between nature and culture, therefore, efforts should be made to reunite and reconcile these terms.

This, in fact, has been the main idea around which the most recent Western scholarly debate on nature has revolved: nature necessarily presupposes culture, which in turn constitutes its framework of analysis and interpretation (cf. Besse 2004).

Despite having widely adopted the opposition between the natural order and the cultural order in his work (and especially in his well-known *Mythologiques*), for instance, Claude Lévi-Strauss was among the firsts to question it. In the preface to the second edition of *Les structures élémentaires de la parenté* (Lévi-Strauss 1967 [1949]), the anthropologist suggested a real "model of transnaturality" (as maintained by Sedda 2021), that is to say, "a translational doubling of nature and the natural, which involves and profoundly alters the statute of culture"[25] (*Ibid.*: 46, my translation). In fact, according to Lévi-Strauss, the distinction between nature and society (or culture) does contain a logic, which fully justifies its use as a methodological tool: "Man is both a biological being and a social individual. Among his responses to external or internal stimuli, some are wholly dependent upon his nature, others upon his social environment" (Lévi-Strauss 1967 [1949]; Engl. Trans. 1969: 3). Although it is not always easy to distinguish between these two dimensions, in the scholar's view, "to deny or to underestimate [their] opposition is to preclude all understanding

---

[23] "Jusqu'à présent ..., quelles que soient la forme prises par leurs relations, la nature était pensée comme première, chronologiquement et ontologiquement, par rapport à la culture. La culture venait après la nature, qui en était pour ainsi dire le cadre".

[24] Whose notes were posthumously collected and published in the volume *La Nature: Notes, cours du Collège de France* (1995).

[25] "Un raddoppiamento traduttivo della natura e del naturale, che coinvolge e stravolge in profondità lo statuto della cultura".

of social phenomena" (*Ibid.*: 4). This led him to prioritise it in his research, where nature emerges as what is constant, universal and spontaneous, as opposed to the normativity, relativity and particularity of culture. However, he problematised the relation between these two poles, referring to culture in terms of a movement of "duplication" of nature, which is "neglectful" of its origin:

> Ultimately we shall perhaps discover that the interrelationship between nature and culture does not favour culture to the extent of being hierarchically superimposed on nature and irreducible to it. Rather it takes the form of a synthetic duplication of mechanisms already in existence but which the animal kingdom shows only in disjointed form and dispersed variously among its members—a duplication, moreover, permitted by the emergence of certain cerebral structures which themselves belong to nature (*Ibid.*: 14).

Culture therefore emerges as "a nature that has (deliberately) forgotten what it was. Or, to use another image, a daughter who disowned her mother"[26] (Sedda 2021: 47, my translation)—which highlights a problematic aspect in its opposition to nature.

A more markedly critical vision of the relation between nature and culture emerged in Bruno Latour's reflections, and in particular in his work *Nous n'avons jamais été modernes* (1991). According to the French scholar, the opposition of these two concepts is in fact the product of a specific ontology, that of Western modernity, which overcame—and at the same time concealed—other ontologies. In his view, the relation between nature and culture is not to be discarded, but rather "pluralised", recognising that, in addition to the dualist ontology (represented by the formula "nature/culture", with a dividing bar in the middle) adopted in Western culture, there is at least another one: a monist ontology which considers the two dimensions inseparable from each other (as reflected by the formula "nature-culture", with a connecting dash in the middle).

Latour also returned on this subject in *Politiques de la nature* (1999), further insisting on a conception of nature as the result of a double construction, both political and scientific. Provocatively depicting it as a "blend of Greek politics, French Cartesianism, and American parks" (Latour 1999; Engl. Trans. 2004: 5), he insisted on the ideological character of this concept, inviting to abandon once for all the idea of an "otherness" separated from the social sphere.

Accordingly, anthropologists Eduardo Viveiros de Castro and Philippe Descola remarked on various occasions that nature is not to be understood as a common background or basis for various cultural formations, as it is generally the case in Western thought. On the contrary, the constitutive *multiplicity of natures* (rigorously in the plural form) characterising our world should be recognised.

More specifically, Viveiros de Castro showed how the general conception of nature and culture is totally subverted in Amerindian thought, according to which culture is related to universality, while nature corresponds to particularity: "'Culture', i.e. the subject, would here be the form of the universal, 'nature', i.e. the object, the form of

[26] "Una natura che ha scordato (volutamente) di essere tale. O per usare un'altra immagine, una figlia che ha ripudiato la propria madre".

the particular" (Viveiros de Castro 1996: 116, my translation).[27] Hence, he suggested a *multinaturalist* approach, envisioning a *perspective* model based on the idea that

> The manifest bodily form of each species is an envelope (a 'clothing') that conceals an internal humanoid form, usually visible to the eyes of only the particular species and of 'transspecific' beings such as shamans. ... If we conceive of humans as somehow composed of a cultural clothing that hides and controls an essentially animal nature, Amazonians have it the other way around: animals have a human, sociocultural inner aspect that is "disguised" by an ostensibly bestial bodily form (Viveiros de Castro 2004: 465).

In this sense, animals and spirits are considered "people", just as humans, and are therefore endowed with a *point of view*, or a "subject position" (which, according to Viveiros de Castro, is located precisely in the body).

> Thus the idea that culture is universal to human beings and distinguishes them from the rest of nature falls apart, as we are faced here with ... a collective in which humans, animals, plants, and even minerals, tools, and astronomical bodies are all agents, and where all of (human) human life, from kinship to politics to medicine, is arranged and conducted accordingly (Viveiros de Castro 2009; Engl. Trans. 2014: 12).

As a result, a multiplicity of natures, each associable with the specific point of view that accesses it, arise, in evident contrast with the Western idea of a multiplicity of cultures emerging from a sole, common nature.

A radical rethinking of the nature/culture dichotomy was also advanced by Descola (2005), who rejected any separation between these two dimensions, rather considering the former, as well as the beings that compose it, "functions" of the latter. Drawing on Michel Foucault's reflections (1966), the French anthropologist focused his attention precisely on the relationships of continuity and discontinuity between nature and culture, with particular reference to what he described as a duality recognised by—and crucial for—all mankind: that between *physicality*, intended as a being's "external form, substance, the physiological, perceptive and sensori-motor processes" (Descola 2005: Engl. Trans. 2013: 116), and *interiority*, or "mental states" (Descola 2014: 275), namely "what we generally call the mind, the soul, or consciousness: intentionality, subjectivity, reflexivity, feelings, and the ability to express oneself and to dream" (Descola 2005; Engl. Trans. 2013: 116). In his view, it is precisely the relation between these two poles that allows humans to emphasise or minimise continuity or difference between themselves and non-humans. As a result, he identified four ontologies—conceived as "systems of the properties of existing beings[, which] serve as a point of reference for contrasting forms of cosmologies, models of social links, and theories of identity and alterity" (*Ibid.*: 121): animism, naturalism, analogism, and totemism (see Fig. 2.1).

[27] "A 'cultura' ou o sujeito seriam aqui a forma do universal, a 'natureza' ou o objeto a forma do particular".

| | | | |
|---|---|---|---|
| *Similar interiorities Dissimilar physicalities* | **Animism** | **Totemism** | *Similar interiorities Similar physicalities* |
| *Dissimilar interiorities Similar physicalities* | **Naturalism** | **Analogism** | *Dissimilar interiorities Dissimilar physicalities* |

**Fig. 2.1** The four ontologies described by Descola

*Animism*, which is mostly spread in South and North America, as well as in Siberia and some parts of Southeast Asia, emphasises physical differences between beings, while recognising that they share the same type of interiority. Identical internal essences, in other terms, are thought to be enclosed in different types of bodies, which—as Viveiros de Castro highlighted (see *supra*)—are conceived as "clothing" that induces contrastive perspectives on the world.

*Totemism* foresees a continuity of both interiority and physicality between humans and non-humans, as exemplified by aboriginal Australian cosmologies. The totem of a group of humans, generally an animal or a plant, is in this case thought to share general attributes of physical conformation and substance, as well as of temperament and behaviour, with all the human and nonhuman beings that are affiliated to it, by virtue of a common origin localised in space.

By contrast, *analogism* extends the discontinuity between humans and non-humans to both interiority and physicality. All beings, in other terms, are seen as "fragmented into a multiplicity of essences, forms, and substances separated by minute intervals, often ordered along a graded scale, such as in the Great Chain of Being, which served as the main cosmological model during the Middle Ages and the Renaissance" (Descola 2014: 276), as well as in many contemporary Asiatic contexts (and especially in China), West Africa, Mesoamerica and the Andes. Such a multiplicity is organised by means of analogies connecting the intrinsic properties of each independent entity, which allows the world to become intelligible and tolerable.

Finally, *naturalism*, which postulates a continuity of physicality, but a discontinuity of interiority, represents the main ontology in Western cultures. In fact, "since Descartes and above all Darwin, we have no hesitation in recognizing that the physical component of our humanity places us in a material continuum within which we do not appear to be unique creatures any more significant than any other organized being" (Descola 2005; Engl. Trans. 2013: 173). However, such a material continuity is related to a discontinuity of interiorities, which marks the superiority of the human over the non-human:

> What, for us, distinguishes humans from nonhumans is the mind, the soul, subjectivity, a moral conscience, language, and so forth, in the same way as human groups are distinguished from one another by a collective internal disposition that used to be called "Volksgeist," or

"génie d'un peuple," but is more familiar to us now under its modern label of "culture" (Descola 2014: 277).

Through the identification of these four systems, Descola effectively showed that not only are there different cultures on the planet, but there are also different natures, that is to say, different modes of perception and conception of the world. In this sense, he overcame the distinction between nature and culture, circumscribing it in space and time and showing that other "ways of identification" between man and the world are possible in the formation of human cultures.

Similar reflections have also appeared in the philosophical field, where several voices have increasingly stood out against the nature/culture (or "natural/artificial") dualism, claiming that it should be dismissed either because everything, including humanity, is artificial (see, for instance, Pitt 2011), or rather because we are ourselves "part of nature" (Callicott 1992: 18), and as such, "everything we do is part of nature, and is natural in that primary sense" (Sober 1986: 180). Other scholars (see, for instance, Casetta 2020) rather suggested to reframe such a distinction in dynamic terms, proposing an interactionist perspective, that is to say, a process of hybridisation between the two poles, by considering both naturalness and artificialness as gradable and relational properties.

The idea of hybridisation has also been particularly stressed in the semiotic field, as exemplified by Gianfranco Marrone's work *Addio alla natura* (2011), which points precisely to "internaturality"[28] as an attempt to find the common forms of all the "natures" arising from the discourses that circulate globally, insisting on the need to consider not only the paradigmatic existence of multiple natures in different cultures—i.e. the "ontologies" described by Descola—, but also, and most importantly, "the syntagmatic presence, and the resulting contrast, of various ontologies within the same culture" (Marrone 2018: 121).

In this view, natures—just like cultures—are not only multiple, but also and necessarily *intertwined* with each other. This means that nature has no pre-established identity at all, but rather emerges as "a constructed evidence, as an artefact that has forgotten the work necessary to produce it, the result of a naturalisation process"[29] (Marrone 2012: 10, my translation). In other words, it is an *effect of meaning*, namely the outcome of a series of *discursive operations*.[30]

Similarly, Claudio Paolucci (2012) urged to focus on the ideologies[31] underlying the notions of "nature" and "culture", rejecting both naturalism and culturalism as they are based on the same distribution of semantic values, and simply favour a

[28] I.e. English translation of the Italian term "internaturalità", as rendered by the author himself (see, for instance, Marrone 2018); the form "internaturalism" has also been used, though less frequently, in some discussions (in English) on this topic (see, for instance, Cravero 2013).

[29] "Un'evidenza costruita, un artefatto che ha dimenticato il lavoro necessario per produrla, l'esito di una procedura di naturalizzazione". In this regard, it is interesting to note the analogy with Roland Barthes' reflections on myth and the processes of naturalisation characterising it (see Barthes 1957 and *supra*).

[30] A particular emphasis on the "discursive value" of nature can also be found in Festi (2012).

[31] Intended in the sense described above (Sect. 1.1), and further discussed in particular by Eco (1968, 1976); cf. Stano and Leone (2023).

different pole in the opposition. In fact, as it clearly emerges from the semiotic square[32] that articulates the relationships between nature and culture (Fig. 2.2), it is evident that the two concepts mutually institute and transform themselves, and hence represent "the momentary and evanescent outcomes of a semiotic articulation on the base of which any substance in the world can be subsumed, so that everything can be now nature and now culture"[33] (Marrone 2012: 11, my translation).

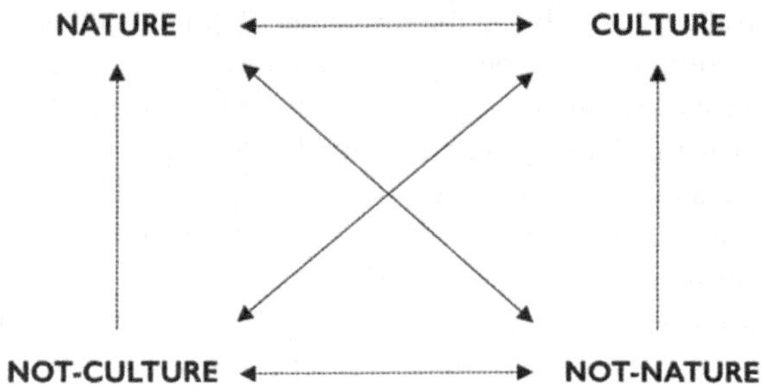


**Fig. 2.2** The semiotic square representing the opposition between nature and culture

The discrimen between nature(s) and culture(s) therefore varies depending on the case, that is to say, as related to the discourses and practices (i.e., in a semiocultural perspective,[34] the "cultures") that "articulate" nature, and its relations to cultural dynamics. Moreover, as highlighted by Dario Martinelli, "it is unacceptable to treat [these two poles] separately, because too many and too complex are the relations between the two" (Martinelli 2010: 58). As Sebeok (1991) also remarked, culture is not in contrast to nature, it is part of it, which makes the very opposition between nature and culture unbearable. Based on such ideas, a particular approach in semiotics, known as "ecosemiotics",[35] has also arisen in the last decades (see in particular Nöth 1996, 1998, 2001; Hoffmeyer 1996a, b; Kull 1998; Maran 2020), in the aim to investigate "nature's structure as it appears, its classification (*syntactics*)" (Kull 1998: 351, my emphasis), as well as its significance for people (*semantics*), and the "personal or social relation to the components of nature, which can be one's participation in nature (pragmatics)" (*Ibidem*).

However, as Festi (2012: 233) lamented, such academic obituaries of nature (both in the semiotic field and in other domains) are still contrasted by its promotional celebration. Moreover, "despite all the theoretical difficulties, as a matter of fact, the natural/artificial distinction is alive and well, not only in people's minds, but also in conservation practices and policies" (Casetta 2020: 4). In sum, the idea of a "pure

[32] I.e. "the visual representation of the logical articulation of any semantic category [in this case, nature *vs.* culture]" (Greimas and Courtés 1979; Engl. Trans. 1982: s.v. "square, semiotic"; cf. Greimas 1966).

[33] "Gli esiti momentanei ed evanescenti di un'articolazione semiotica a partire da cui qualsiasi sostanza del mondo può essere sussunta, per cui ogni cosa può essere ora natura ora cultura".

[34] See in particular Lotman (1984, 1990).

[35] Which also intersects other branches and approaches, such as biosemiotics and zoosemiotics.

Nature", defined in opposition to any form of cultural transformation, seems to persist in a number of texts, discourses and practices that circulate throughout contemporary Western cultures, regardless of the warnings and new approaches considered above. A critical study of such texts, discourses and practices—which will be precisely the object of the following chapters—, and of the processes of naturalisation they entail, is therefore more necessary than ever.

## References


Agrippa von Nettesheim, Heinrich Cornelius. 1533. *De occulta philosophia libri tres*, ed. Vittoria Perrone Compagni. 1992. Leiden: Brill.

Arias-Maldonado, Manuel. 2015. What is nature? In Id. *Environment and Society. Socionatural Relations in the Anthropocene*, 17–32. Cham: Springer.

Aristotle. *Φυσικὴ ἀκρόασις*. 4th cent. BCE. English Trans. Reeve, Charles D.C. 2018. *Physics*. Cambridge, MA: Hackett Publishing Company.

Aristotle. *τὰ μετὰ τὰ φυσικά*. 4th cent. BCE. English Trans. Sachs, Joe. 2002 [1992]. *Metaphysics*. Santa Fe, NM: Green Lion Press.

Augustine of Hippo. 413–426. *De civitate Dei (contra paganos)*. English Trans. Babcock, William, with notes by Ramsey, Boniface. 2012. *The City of God*. Hyde Park, NY: New City Press.

Bacon, Sir Francis. 1620. Novum organum. Originally published in *Instauratio Magna*, 2nd part. London: Joannem Billium Typographum Regium. New edition by Joseph Devey. 1902. New York, NY: P.F. Collier & Son.

Barthes, Roland. 1957. *Mythologies*. Paris: Editions de Seuil. English Trans. Lavers, Annette. 1972. *Mythologies*. New York, NY: The Noonday Press—Farrar, Straus and Giroux.

Benveniste, Émile. 1948. *Noms d'agent et noms d'action en indo-européen*. Paris: Adrien-Maisonneuve.

Besse, Jean-Marc. 2004. Nature et culture. *HyperGeo—Encyclopédie électronique consacrée à l'épistémologie de la géographie*. https://hypergeo.eu/nature-et-culture/. Accessed 21 Dec 2022.

Buffon, Georges Louis Leclerc, comte de. 1770. *Histoire naturelle, générale et particulière, avec la description du Cabinet du Roy*. Paris: Imprimerie Royale.

Callicott, J. Baird. 1992. La nature est morte, vive la nature! *The Hastings Center Report* 22 (5): 16–23.

Callicott, J. Baird, and Roger T. Ames. 1989. Introduction: The Asian traditions as a conceptual resource for environmental philosophy. In *Nature in Asian Traditions of Thought: Essays in Environmental Philosophy*, ed. J. Baird Callicott and Roger T. Ames, 1–21. New York, NY: SUNY Press.

Campanella, Tommaso. 1620. *De sensu rerum et magia*. Frankfurt.

Casetta, Elena. 2020. Making sense of nature conservation after the end of nature. *History and Philosophy of the Life Sciences* 42: article number 18.

Caspers, Christiaan L. 2010. φύω. In *Lexikon des frühgriechischen Epos. Begründet von Bruno Snell. Im Auftrag der Akademie der Wissenschaften in Göttingen vorbereitet und herausgegeben vom THESAURUS LINGUAE GRAECAE*, vol. 4: *P–Ω*, 1068–1072. Göttingen: Vandenhoeck & Ruprecht.

Castree, Noel. 2014. *Making Sense of Nature*. London and New York, NY: Routledge.

Cravero, Claudio. 2013. Internaturalism: Towards an hybridisation of human and non-human. In AA.VV., *Internaturalità/Internaturalism*, 42–44. Torino: PAV Art Program.

Curley, Edwin, ed. and trans. 1985. *The Collected Works of Spinoza*, vol. 1. Princeton, NJ: Princeton University Press.

D'Alembert, Jean le Ronde, et al. 1765. Nature. In *Encyclopédie ou Dictionnaire raisonné des sciences, des arts et des métiers*, vol. 11, ed. Denis Diderot and Jean le Ronde d'Alembert, s.v. "Nature". Paris: Le Breton.

Diderot, Denis. 1753. *Encyclopédie ou Dictionnaire raisonné des sciences, des arts et des métiers*, vol. 3, ed. Denis Diderot and Jean le Ronde d'Alembert. Paris: Briasson—Le Breton—Durand.

Descartes, René. 1664. *Le Monde, ou Traité de la lumière et des autres principaux objets des Senses*. Paris: Michel Bobin & Nicolas le Gras. English Trans. Mahoney, Michael S. 1979. *The World, or Treatise on the Light*. New York, NY: Abaris Books.

Descola, Philippe. 2005. *Par-delà nature et culture*. Paris: Gallimard. English Trans. Lloyd, Janet. 2013. *Beyond Nature and Culture*. Chicago, IL: The University of Chicago Press.

Descola, Philippe. 2011. *L'écologie des autres. L'anthropologie et la question de la nature*. Paris: Éditions Quae. English Trans. Godbout, Geneviève, and Benjamin P. Luley. 2013. *The Ecology of Others*. Chicago, IL: Prickly Paradigm Press.

Descola, Philippe. 2014. Modes of being and forms of predication. *Hau: Journal of Ethnographic Theory* 4 (1): 271–280.

Ducarme, Frédéric, and Denis Couvet. 2020. What does 'nature' mean? *Palgrave Communications* 6 (1): 1–8.

Dunshirn, Alfred. 2019. Physis [English version]. In *Online Encyclopedia Philosophy of Nature/ Online Lexikon Naturphilosophie*, ed. Thomas Kirchhoff. https://journals.ub.uni-heidelberg.de/index.php/oepn/article/view/66404/59195. Accessed 11 Jan 2023.

Eco, Umberto. 1968. *La struttura assente*. Milan: Bompiani.

Eco, Umberto. 1976. *A Theory of Semiotics*. Bloomington, IN: Indiana University Press.

Festi, Giacomo. 2012. Gli oggetti culturali naturali. Fermentazioni traduttive e applicazioni al mondo del vino. In *Semiotica della natura (Natura della semiotica)*, ed. Gianfranco Marrone, 233–255. Milan-Udine: Mimesis.

Ficino, Marsilio. 1484. *Sopra lo amore ovvero Convito di Platone*. English Trans. Sears, Jayne. 1985. *Commentary on Plato's Symposium on Love*. Dallas, TX: Spring Publications.

Ficino, Marsilio. 1489. *De vita libri tres* (or *De triplici vita*). English Trans. Kaske, Carol V., and John R. Clarke. 2002. *Three Books on Life*. Tempe: The Renaissance Society of America.

Frisk, Hjalmar. 2006. *Griechisches etymologisches Wörterbuch*, vol. II: *Κρ–Ω. Vierte, unveränderte Auflage*. Heidelberg: Winter.

Foucault, Michel. 1966. *Les mots et les choses: Une archéologie des sciences humaines*. Paris: Gallimard. English Trans. Sheridan Smith, Alan M. 1971. *The Order of Things: An Archeology of the Human Sciences*. New York, NY: Pantheon Books.

von Goethe, Johann Wolfgang. 1808. *Faust*, ed. Albrecht Schöne. Frankfurt am Main: Deutscher Klassikerverlag, 2005. English Trans. Bayard, Taylor. 1889. *Faust*. London: Frederick Warne & Co. Ltd.

Greimas, Algirdas J. 1966. *Sémantique structurale*. Paris: Larousse.

Greimas, Algirdas J., and Joseph Courtés. 1979. *Sémiotique: Dictionnaire raisonné de la théorie du langage*. Paris: Hachette. English Trans. Crist, Larry, Daniel Patte, James Lee, Edward McMahon II, Gary Phillips, and Michael Rengstorf. 1982. *Semiotics and Language: An Analytical Language*. Bloomington, IN: Indiana University Press.

Grierson, David. 2009. The shift from a mechanistic to an ecological paradigm. *International Journal of Environmental, Cultural, Economic and Social Sustainability* 5 (5): 197–206.

Habgood, John S. 2002. *The Concept of Nature*. Darton: Longman & Todd.

Hadot, Pierre. 2004. *Le voile d'Isis : essai sur l'histoire de l'idée de nature*. Paris: Gallimard. English Trans. Chase, Michael. 2006. *The Veil of Isis: An Essay on the History of the Idea of Nature*. Cambridge, MA, and London, UK: The Belknap Press of Harvard University Press.

Heidegger, Martin. 1976 [1939]. Vom Wesen und Begriff der Φύσις. Aristoteles, Physik B, 1. In *Wegmarken 1919–1958*, vol. 9, ed. Fridrich-Whilhelm von Hermann, 239–301. Frankfurt am Main: Vittorio Klostermann. English Trans. Sheehan, Thomas J. 1976. On the Being and Conception of Φύσις in Aristotle's Physics B, 1. *Man and World* 9 (3): 219–270.

Hoffmeyer, Jesper. 1996a. *Signs of Meaning in the Universe*. Bloomington: Indiana University Press.

Hoffmeyer, Jesper. 1996b. Für eine semiotisch reformulierte Naturwissenschaft. *Zeitschrift Für Semiotik* 18 (1): 31–34.

Homer. 8th cent. BCE. *'Οδύσσεια*. English Trans. Fitzgerald, Robert. 1998. *The Odyssey*. New York, NY: Farrar, Straus and Giroux.
Jaspers, Karl. 1974. *Spinoza—Great Philosophers*, vol. II, ed. Hannah Arendt, trans. Ralph Manheim. New York, NY, and London: Harvest Books.
Kodalak, Gökhan. 2020. Spinoza and architecture: The air of the future. *The Log* 49: 122–145.
Kull, Kalevi. 1998. Semiotic ecology: Different natures in the semiosphere. *Sign Systems Studies* 26: 344–371.
Lammer, Andreas. 2015. Defining nature: From Aristotle to Philoponus to Avicenna. In *Aristotle and the Arabic Tradition*, ed. Ahmed Alwishah and Josh Hayes, 121–142. Cambridge: Cambridge University Press.
Larrère, Catherine, and Raphaël Larrère. 2015. *Penser et agir avec la nature. Une enquête philosophique*. Paris: La Découverte.
Latour, Bruno. 1991. *Nous n'avons jamais été modernes: Essai d'anthropologie symétrique*. Paris: La Découverte. English Trans. Porter, Catherine. 1993. *We Have Never Been Modern*. Cambridge, MA: Harvard University Press.
Latour, Bruno. 1999. *Politiques de la nature. Comment faire entrer les sciences en démocratie*. Paris: La Découverte. English Trans. Porter, Catherine. 2004. *Politics of Nature: How to Bring the Sciences into Democracy*. Cambridge, MA: Harvard University Press.
Lenoble, Robert. 1969. *Histoire de l'idée de nature*. Paris: Albin Michel.
Leone, Massimo. 2012. Nasi possibili: Internaturalità e figurazione. In *Semiotica della natura (Natura della semiotica)*, ed. Gianfranco Marrone, 183–207. Milan-Udine: Mimesis.
Lévi-Strauss, Claude. 1967 [1949]. *Les structures élémentaires de la parenté*, revised ed. Paris-La Haye: Mouton. English Trans. Harle Bell, James, John Richard von Sturmer, and Rodney Needham. 1969. *The Elementary Structures of Kinship*. Boston, MA: Beacon Press.
Licata, Geatano. 2019. Definizione e significati della φύσις in Aristotele a partire dal confronto con la τέχνη. Il λόγος come apertura della φύσις alla τέχνη in Fisica B 1–2. In *Natura della tecnica e tecnica della natura*, ed. Rosaria Caldarone, 11–29. Brescia: Morcelliana.
Ljungberg, Christina. 2001. Wilderness from an ecosemiotic perspective. *Signs Systems Studies* 29 (1): 169–186.
Lotman, Jurij M. 1967. К проблеме типологии культуры. *Труды По Знаковым Системам (signs Systems Studies)* 3 (1): 30–38.
Lotman, Jurij M. 1984. О семиосфере. *Труды по знаковым системам (Signs Systems Studies)* 17: 5–23. English Trans. Clark, Wilma. 2005. On the semiosphere. *Signs Systems Studies* 33 (1): 205–229.
Lotman, Jurij M. 1990. *Universe of the Mind: A Semiotic Theory of Culture*. Bloomington, IN: Indiana University Press.
Lucretius (Titus Lucretius Carus). 50 BCE. *De rerum natura*. English Trans. Rouse, William Henry, revised by Ferguson Smith, Martin. 1992 [1924]. *On the Nature of Things*. Cambridge, MA: Harvard University Press.
Maran, Timo. 2020. *Ecosemiotics: The Study of Signs in Changing Ecologies*. Cambridge, MA: Cambridge University Press.
Marrone, Gianfranco. 2011. *Addio alla natura*. Torino: Einaudi.
Marrone, Gianfranco, ed. 2012. *Semiotica della natura (Natura della semiotica)*. Milan-Udine: Mimesis.
Marrone, Gianfranco. 2015. La nature au supermarché. Sur le packaging des produits dits biologiques. *Actes Semiotiques* 118: 1–14. https://www.unilim.fr/actes-semiotiques/5379&file=1. Accessed 14 Dec 2022.
Marrone, Gianfranco. 2018. Bestiality: Animal cultures. In *Semiotics of Animals in Culture*, ed. Gianfranco Marrone and Dario Mangano, 121–134. Cham: Springer.
Martinelli, Dario. 2010. *A Critical Companion to Zoosemiotics: People, Paths, Ideas*. Dordrecht-Heidelberg-London-New York: Springer.
Merchant, Carolyn. 1980. *The Death of Nature: Women, Ecology, and the Scientific Revolution*. San Francisco, CA: Harper and Row.

Merchant, Carolyn. 2003. *Reinventing Eden: The Fate of Nature in Western Culture*. New York, NY, and London: Routledge
Merchant, Carolyn. 2005 [1992]. *Radical Ecology: The Search for a Livable World*. New York, NY, and London: Routledge.
Merleau-Ponty, Maurice. 1995. *La Nature: Notes, cours du Collège de France (1956–1960)*. Paris: Seuil. English Trans. Vallier, Robert, with notes by Séglard, Dominique. 2003. *Nature: Course Notes from the Collège de France*. Evanston, IL: Northwestern University Press.
Mill, John Stuart. 1904 [1874]. On Nature. In *Nature, the Utility of Religion, and Theism*, 373–402. London: Longmans, Green & Co. E-version by the Philosophy Department at Lancaster University. https://www.marxists.org/reference/archive/mill-john-stuart/1874/nature.htm. Accessed 19 Jan 2023.
Naddaf, Gerard. 2005. *The Greek Concept of Nature*. Albany, NY: State University of New York Press.
Nicholas of Cusa. 1440. *De docta ignorantia*, 3 books. English Trans. Hopkins, Jasper. 1981. *On Learned Ignorance*. Minneapolis, MN: The Arthur J. Banning Press.
Nöth, Winfried. 1996. Ökosemiotik. *Zeitschrift Für Semiotik* 18 (1): 7–18.
Nöth, Winfried. 1998. Ecosemiotics. *Sign Systems Studies* 26: 332–343.
Nöth, Winfried. 2001. Ecosemiotics and the semiotics of nature. *Sign Systems Studies* 29 (1): 71–81.
Oehlschläger, Max. 1991. *The Idea of Wilderness: From Prehistory to the Age Ecology*. New Haven, CT, and London, UK: Yale University Press.
Owens, Joseph. 1968. Teleology of nature in Aristotle. *The Monist* 52 (2): 159–173.
Paolucci, Claudio. 2012. *Physis* e *nomos*. Ideologie della natura tra catarsi, empatia e percezione sessuale. In *Semiotica della natura (Natura della semiotica)*, ed. Gianfranco Marrone, 79–102. Milan-Udine: Mimesis.
Passmore, John Arthur. 1974. *Man's Responsibility for Nature: Ecological Problems and Western Tradition*. New York, NY: Scribner.
Pellicer, André. 1966. Natura, Étude sémantique et historique du mot latin. *Revue Belge De Philologie Et D'histoire* 47 (3): 978–982.
Pepper, David. 1996. *Modern Environmentalism. An Introduction*. London, UK, and New York, NY: Routledge.
Pitt, Joseph C. 2011. *Doing Philosophy of Technology: Essays in a Pragmatist Spirit*. Dordrecht: Springer.
Sebeok, Thomas A. 1991. Communication. In *A Sign is Just a Sign*, 22–35. Bloomington, IN: Indiana University Press.
Sedda, Franciscu. 2021. Nello specchio dell'antropologia: La natura, la cultura, il semiotico. *Estudos Semióticos* 17 (2): 44–67.
Simberloff, Daniel. 2014. The "balance of nature"—Evolution of a panchreston. *PLoS Biology* 12 (10): e1001963.
Sober, Elliott. 1986. Philosophical problems for environmentalism. In *The Preservation of Species: The Value of Biological Diversity*, ed. Bryan G. Norton, 173–194. Princeton, NJ: Princeton University Press.
Soper, Kate. 1995. *What is Nature? Culture, Politics and the Non-Human*. London: Blackwell.
Spinoza, Baruch (Benedictus de). 1994. *A Spinoza Reader: The Ethics and other works*, ed. and trans. Edwin Curley. Princeton, NJ: Princeton University Press.
Stano, Simona, and Massimo Leone, eds. 2023. *Ideologia—Ideology. Special issue of Lexia*, 41–42. Rome: Aracne.
Telesio, Bernardino. 1971 [1565–86]. *De rerum natura iuxta propria principia*, ed. Cesare Vasoli. Hildesheim–New York: Olms.
Viveiros de Castro, Eduardo. 1996. Os pronomes cosmológicos e perspectivismo ameríndio. *Mana* 2 (2): 115–144.

Viveiros de Castro, Eduardo. 2004. Exchanging perspectives: The transformation of objects into subjects in amerindian ontologies. *Common Knowledge* 10 (3): 463–484.
Viveiros de Castro, Eduardo. 2009. *Métaphysiques cannibales*. Paris: PUF. English Trans. Skafish, Peter. 2014. *Cannibal Metaphysics: For a Post-Structural Anthropology*. Minneapolis, MN: Univocal.
Vogel, Steven. 2015. *Thinking like a Mall: Environmental Philosophy after the End of Nature*. Cambridge, MA: The MIT Press.

# Chapter 3
# Mother Nature: Representations, Isotopies, and Meanings

**Abstract** This chapter deals with the analysis of Mother Nature, intended as a personification of nature, whose valorisation and figurativisation has evidently changed over time. After the description of the main ideas and features that have been associated with Mother Nature along time, its Western iconography is described moving from the past to the present, and paying particular attention to the main isotopies (intended, in a Greimassian perspective, as the repetition, along a syntagmatic chain, of traits ensuring homogeneity to a discourse-enunciated) and meanings characterising contemporary imaginaries. This also allows highlighting crucial aspects for the development of the following chapters, which will build precisely on some of the elements discussed in these pages to enhance the reflection on the current mythologies of nature.

## 3.1 Mother, Female, Goddess: Who is Mother Nature?

Also referred to as "Mother Earth" or "Earth-Mother", "Mother Nature" is a common personification of nature, considered the "source and guiding force of creation" (Merriam-Webster 2022: s.v. "Mother Nature"). In fact, as illustrated in the previous chapter, the very etymology of the word *nature* (derived from the Latin verb *nasci*, which translates the Greek *phýein*, "to be generated") refers to the birth of all things and their development according to their own principles (cf. Brevini 2013: 29).

As a generating source, Nature has been traditionally associated with the female sphere and, more precisely, with motherhood.[1] In his *Naturalis Historia* (Book XXXVII), for instance, Pliny the Elder addressed "her" as the "parent of all things". Accordingly, as Hadot (2004) observed, Mesomedes defined Nature as the "principle and origin of all things, Ancient mother of the world" (*Hymn to Nature*; Engl. Trans. in Hadot 2004: 27; cf. Smolak 1987), and Marcus Aurelius invocated her by using the words "Everything is fruit to me which thy seasons bring, O Nature: from thee

[1] For more recent reflections on the association of nature with the female sphere and motherhood, see in particular Merchant (1980), Toland Frith (1995).

 
S. Stano, *Critique of Pure Nature*, Numanities - Arts and Humanities in Progress 26,
https://doi.org/10.1007/978-3-031-45075-4_3

are all things, in thee are all things, to thee all things return" (*Meditations* 4.23; Engl. Trans. in Long 1880 [1862]: 98).

Moreover, such an idea of Nature has been conventionally connected with the divine sphere:

> She was thought to exhale the breath of life, which nourished living organisms on her surface… fluids flowed within her and the water came out of her springs like blood. Within her body there were veins, some of which contained liquids and other solidified fluids. She bore stones within her womb and nurtured them as they grew, like embryos, within her, ripening at their own slow pace (Sheldrake 1991: 15).

Actually, the first written reference to Mother Earth or Mother Goddess was recorded in the early 7th century BCE by the Greek poet Hesiod in his *Theogony*, where Gaia (from Ancient Greek *Γαῖα*, a poetical form of *Γῆ*, *Gē*, meaning "earth", cf. Liddell and Scott 1940: s.v. "γαῖα") played a crucial role in the birth of the universe—intended, as highlighted by Trzaskoma et al. (2016: 129), as both the origin of the gods (i.e. a "theogony", as the title of Hesiod's work itself suggests) and the creation of the cosmos (i.e. a cosmology).[2] Described as a primal deity, Nature was revered as the Mother of all gods and living creatures. Such an idea was further reinforced over time, as confirmed, for instance, by Mesomedes of Crete (2nd century), who also described nature as the principle and origin of everything, the "Ancient mother of the world" (see Smolak 1987). Accordingly, an orphic hymn from the same period celebrated "Mother Goddess" (see Quandt 1995; Engl. Trans. in Raine and Harper 1969: 221–223), and various other texts confirmed this view, either in a general way or with a specific reference to Gaia.[3] In fact, the association of such a generating principle with specific goddesses was not unusual, as demonstrated by the importance of Tiamat in the Mesopotamian culture, the cult of Isis in Ancient Egypt (and even in more recent times, cf. Hadot 2004), that of Tellus (Mater) in the Roman Empire, and the worship of Nana, Ninhursag and Mami in other ancient cultures[4] (see Murdock 1990).

Towards the fifth century, nature took on a more mysterious characterisation, becoming mostly inaccessible (which, as it will be highlighted in the following paragraph, had a great impact on the representation of Mother Nature). Moreover, as

---

[2] In fact, according to the verses of the poem, "In the beginning there was only Chaos, the Abyss/ But then Gaia, the Earth, came into being/Her broad bosom the ever-firm foundation of all" (*Theog.* 116–119, Engl. Trans. in Trzaskoma et al. 2016: 135). The goddess then gave birth to Ouranos (the "starry Heaven"), and by their union more offspring ("Ocean with its deep currents/And also: Coios, Crios, Hyperion, Iapetos/Theia, Rheia, Themis, Mnemosyne/Gold-crowned Phoibe, and lovely Tethys", *Theog.* 133–136, Engl. Trans. in Trzaskoma et al. 2016: 135) also came, with a fundamental impact on the generation of the physical world too.

[3] The Anatolian goddess Cybele was also sometimes partially assimilated to aspects of the Earth-goddess Gaia, as well of her possibly Minoan equivalent Rhea and of the harvest-mother goddess Demeter. However, she was an ambivalent deity, symbolising both the creative and the destructive force of Nature (see Roller 1999).

[4] Also including still worshipped deities such as Kali in Hindu contexts (though with multiple variations and valorisations, see Harding 1993) and Pachamama, or Mama Pacha, in the Andes (see Dransart 1992).

discussed in Chap. 1, it progressively lost its divine character, being conceived itself as a creation of God—to which it was therefore subordinated, though maintaining a certain autonomy.

After a partial return to the classical imaginary during the Renaissance period, the so-called Scientific Revolution eventually transformed nature from *Terra Mater* into a simple "source" of raw materials, thus totally disabling any divine characterisation, as well as most ethical and cognitive constraints against the manipulation and exploitation of nature in the name of science, technology and development: "Mother Nature was no more than dead matter, moving in unfailing obedience to God-given mathematical laws" (Sheldrake 1991: 49).

The idea of Mother Nature thus basically disappeared from Western thought, or better—as the following paragraphs will illustrate—moved farther and farther into the background, making it become an extremely variable figure to be adapted to different contexts and invested with multiple values.

## 3.2 The Iconography of Mother Nature in the Past: Relevant Case Studies in the Western Arts[5]

One of the oldest representations of the idea of Mother Nature can be found in the so-called Venus of Willendorf (Fig. 3.1), which was discovered in the homonymous village in Austria and is estimated to have been made in Palaeolithic times, between 26,000 and 20,000 BCE. Like other archaeological findings from around the prehistoric world (the so-called "Venus figurines"), this sculpture represents a voluptuous aniconic female figure, with large breasts and a considerable stomach that extends over an accentuated pubic area, recalling the ideas of fertility and nurturing.[6]

However, as highlighted in the previous paragraph, the first direct reference to Mother Earth only appeared in the seventh century BCE, as related to the Greek goddess Gaia, whose representations are manifold. As Collignon (1890) illustrated, the deity was generally represented seated, as she appeared in a statue seen by Pausanias in the sanctuary of Demeter at Parae, as well as in vase painting (see Panofka 1842) and bas-reliefs, where she was generally depicted as a matronly woman, wearing a *polos* crown with a long veil, and handing her new-born Erichthonius to Athena. Sometimes she also appeared half length, with her bust rising out of the ground and a wealth of flowing hair, accompanied by one or more of her children. In any case, her maternal character was particularly emphasised, sometimes also

[5] The attention is put particularly on the Western semiosphere in accordance with the main scope and perspective adopted in this book, as described in Chap. 1.

[6] A number of variants of such a figure exist, also with evident differences related to geographical and temporal contexts. Provided the purposes of this chapter, only this prototypical artefact and the general idea associated with it are considered. For further details on each variant, please refer to the rich archaeological literature on this topic.

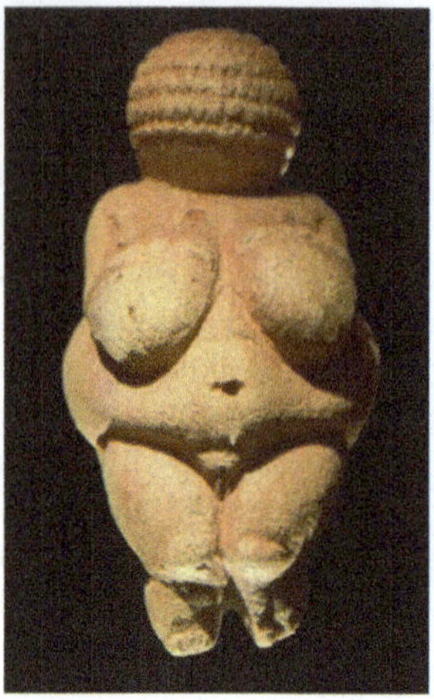

**Fig. 3.1** The Venus of Willendorf, c. 26,000 BCE, Naturhistorisches Museum (Austria), picture by Matthias Kabel[7]

making her become a secondary character, with visual priority given to her sons (especially when Erichthonius or the Titans were represented).

As mentioned above, another divine figure generally personifying Nature was Isis, mostly identified with Artemis—or, in the Roman culture, with Diana of Ephesus (Hadot 2004). According to Macrobius' *Saturnalia*, Isis was worshipped "in a kindred cult as either earth or the natural order subject to the sun. That is why the goddess' entire body is thick with an unbroken series of breasts: the whole of the earth or natural order gains its sustenance from her" (*Saturnalia—Book I*, 20.18).

In fact, as Pierre Hadot (2004) highlighted in his book *Le voile d'Isis: essai sur l'histoire de l'idée de nature*, most representations of Isis as a personification of Nature, up until the 15th century, portrayed her as a naked woman, to symbolise her simplicity and transcendence (see also Gravelot and Cochin 1791: s.v. "Natura"; Kempt 1973: 19), and with a number of breasts, to emphasise the idea of nurturing and motherhood. Since the 16th century the influence of the iconography of Artemis became more visible, with a crown (such as in the famous marble stand *The Goddess of Nature* by Niccolò Tribolo, c. 1529) or a veil (such as in *Diana of Ephesus as allegory of Nature* by Joseph Werner, represented in Fig. 3.2) covering the female figure's head and falling down on her body, leaving her multiple breasts visible.

Sometimes a vulture also accompanied the divine female figure (such as in the representation included in the *Emblemata cum aliquot nummis antiqui operis* by Joannes Sambucus (1564: 74), known as *Physicae ac Metaphysicae differentia*), possibly designating an Egyptian heritage and symbolising the ideas of purification and rebirth associated with nature (see Ripa 1593; Kempt 1973).

The importance of the veil further increased over time, also independently from the figure of Isis, becoming a symbol of the secrets and inaccessibility of Nature. However, with the development of science and technology between the 17th and the

---

[7] Creative Commons Attribution-Share Alike 3.0 Unported license, https://commons.wikimedia.org/wiki/File:Venus_of_Willendorf_frontview_retouched_2.jpg (Accessed: 7 December 2022).

**Fig. 3.2** *Diana of Ephesus as allegory of Nature*, Joseph Werner (c. 1680), Art Institute Chicago (United States)

18th centuries, a marked process of "unveiling" began. Hadot (2004: 233) provides evidence of this trend by recalling the frontispiece of Gerhard Blasius' *Anatome animalium* (1681, represented in Fig. 3.3), where Science is portrayed as a young woman with a flame (i.e. a symbol of the desire for knowledge) on her head, and a lens and a scalpel (symbols of science and medicine) in her hands, revealing a woman with four breasts and the representation of the seven planets among them.

**Fig. 3.3** Frontispiece of *Anatome animalium* (1681)

Surrounded by animals and with her right arm holding a sceptre, on which a vulture perches, this figure clearly recalls the first representations of Nature, as described above. However, both the lifting of her veil by Science and the presence of two *putti* dissecting an animal (one) and examining its entrails (the other) symbolise the crucial role played by scientific work in approaching and getting to know nature.[8]

Similarly, as Hadot remarks, the frontispiece of the book *De la Nature et des ses Lois* by François Peyrard (1793), *La philosophie de la Nature* (Fig. 3.4), recurred to an unveiling act (in this case performed by the personification of Philosophy) as a metaphor of the triumph of illuminist reasoning over the idea of a transcendental view of Nature, intended as a superior and inaccessible entity. Nature, in other terms, is "laid bare", hence becoming accessible in its "essence" through reasoning and "culture".

Thus, the figure of "Mother Nature" progressively blurred, and was further weakened by Romanticists, which promoted a sentiment of both anguish and amazement[9] towards nature, mainly identifying it with the figure of the *sphinx*—i.e. a wild beast with the bust of a beautiful young woman, whose duplicity reflected the ambivalence

**Fig. 3.4** Frontispiece of *De la Nature et des ses Lois* (1793)

[8] In accordance with what stated in Chap. 1, the lower case is used here to highlight the passage from a vision of Nature as a deity and supreme—and thus inaccessible—force to an object of study, which humanity can reach through Science (which, by contrast requires here a capital initial).

[9] In his *Dialogue Between Nature and an Icelander*, for instance, Leopardi (1835 [1824]) described nature as "an enormous woman … seated on the ground, resting her back against a mountain", whose "countenance [was] both magnificent and terrible" (Engl. Trans. 1882: 73).

of nature between beauty and ferocity, and the feeling of both wonder and horror it can provoke in man. This progressively led to the awareness that humanity itself is part of nature (see Hadot 2004: 314), fostering a process of resemantisation of nature—which, as I will show in detail in the following paragraphs, has determined relevant changes in the figure of Mother Nature, passing through multiple figures and meanings.

## 3.3 The Iconography of Mother Nature in the Present: Relevant Case Studies in Western Popular Culture

Despite the above-described process of progressive loss of importance of Mother Nature, representations of this figure are still abundant in Western contemporary imaginaries. The following paragraphs will analyse some of the most significant ones,[10] ranging from interesting examples in music to visual texts (e.g. comics and print advertisements) and audiovisual materials (i.e. movies, TV programmes and commercials), in order to point out their main features and effects of meaning.

### *3.3.1 Mother Nature in Music*

In 1971 the American singer-songwriter and guitarist Don McLean released his famous album *American Pie*, reissued in 2003 with two bonus tracks, including the song "Mother Nature"—whose lyrics are reported below:

I want her so bad

Mother Nature has a hold on me

[10] In order to define a coherent corpus of analysis, only representations explicitly referred to as Mother Nature (or Mother Earth) are taken into consideration in this section, therefore excluding figures such as Poison Ivy (in *Batman*, as related to both comic books and other media) and others which, while certainly interesting and recalling features typical of Mother Nature, play a different role in the narratives they are part of. Likewise, the analysis does not involve texts where Mother Nature is recalled but does not find a clear personification, as it happens with Eywa, also known as the "Great Mother" or "All-Mother", in the movie *Avatar* (Cameron 2009) and its sequel *Avatar: The Way of Water* (Cameron 2022). The only known deity of the Na'vi, Eywa is presented as the biological sentient guiding force of life and is believed to act to keep the ecosystem of Pandora in equilibrium. It is not seen as an all-powerful, world-creating deity, but rather as a system connecting all living beings (cf. Izzo 2022). While particular entities, such as the Tree of Souls (which allows the Na'vi to connect their tails to its branches, and therefore to each other and to Eywa itself), or individuals, such as Mo'at (the Tsahìk, or spiritual leader, of the Omatikaya clan, who acts as a "bridge" to Eywa), allow a connection to the Great Mother, this does not find a clear personification in the movie. Finally, other representations, such as Te Fiti in the computer-animated musical fantasy action-adventure film *Moana* (also known as *Vaiana* or *Oceania* in some countries) directed by John Musker and Ron Clements (2006), are not considered because they refer to other cultural imaginaries.

I want her so bad

Mother Nature won't you let me be untied?

'Cause it hurts my pride

To be tossed off like the morning covers

And crossed off, like her other lovers.

Casually

I see her walking by my window—mhmmhmmhmmhmm!

It seems to me I know her well.

But like the flowers in the spring-time,

Growing toward the sunshine,

Her beauty falls upon me in a fragrant spring-time spell

And I want her so ba-aa-aad

Mother Nature has a hold on me

I want her so bad

Mother Nature won't you let me be untied?

'Cause it hurts my pride

To be tossed off like the morning covers

And crossed off, like her other lovers.

Casually

Is she an icy winter woman? Mm-hhmm!

That chills my body when she's near

Sweet fever in the morning

I don't know if I'll survive.

She walks by me as though she didn't

Know I was alive

And I want her so ba-aa-aad

Mother Nature has a hold on me

I want her so bad

Mother Nature won't you let me be untied?

'Cause it hurts my pride

To be tossed off like the morning covers

And crossed off, like her other lovers

Tossed off like the morning covers,

Crossed off, like her other lovers

Tossed off like the morning covers,

Crossed off, like her other lovers.

These lines open the 5′ 01″ song, preceding a long instrumental part, and depicting a particular image of Mother Nature, whose characterisation makes the singer inescapably fall in love with her (as particularly remarked by the verses: "Her beauty falls upon me in a fragrant spring-time spell/And I want her so ba-aa-aad", "has a hold on me", "I want her so bad", and "Mother Nature won't you let me be untied?"—with an *embrayage*, or "engagement",[11] interestingly used in the refrain to simulate a direct call to Mother Nature), though she does not correspond his love. The figure of an "icy winter woman" is in fact used to describe her, also insisting on her perilousness ("That chills my body when she's near/Sweet fever in the morning/I don't know if I'll survive"). A Romantic view of nature therefore seems to find a certain echo in this text, recalling the figure of Terra Mater to emphasise her ambivalent characterisation, both as a source of wonder and pleasure, and as a reason for hurt and sorrow.

A few years after the release of this song, another one, "Death of Mother Nature Suite" (1974) by the progressive rock band Kansas, was produced.

Every day she gets a little weaker

The beauty she once knew has come and gone

We've murdered all her sons and all her daughters

The blood is on your hands, the time has come

And now she's gonna die.

We've strangled all her trees and starved her creatures

There's poison in the sea and in the air

But worst of all we've learned to live without her

We've lost the very meaning of our lives

And now she's gonna die.

Once she ruled the earth with love and wisdom

But we were much too smart to live her way

With greed and lust we tried to rise above her

The ignorance of man will reach an end

---

[11] I.e. "the effect of a return to the enunciation … produced by the suspension of the opposition between certain terms belonging to the categories of actor and/or of space and/or of time, as well as by a negation of the domain of the utterance" (Greimas and Courtés 1979; Engl. Trans. 1982: s.v. "engagement").

'Cause now we're gonna die.

As the above-reported title and lyrics evidently show, Mother Nature is in this case depicted as a dying figure, which is getting weaker and weaker, after having lost her beauty, as well as "all her sons and all her daughters". Accordingly, the text does not address her directly, rather opting exclusively for third-person pronouns and adjectives to mark a "disengagement" (or *débrayage*[12]). Conversely, an *embrayage* (i.e. the use of "we" and "us") is used to involve the listeners and stress everyone's (i.e. men's) role in her death. In fact, the song was meant as a protest against industrialisation, which highlights a conception of nature primarily in negative terms, that is to say, in opposition to "culture", intended as the action of men —who are deemed responsible for having "learned to live without [nature]", thus losing "the very meaning of [their] lives".

### 3.3.2 *Mother Nature in the Mass Media: Comics, Movies and TV Programmes*

Mother Nature is featured in various mass media discourses, and especially in comics, movies and TV programmes. The case of *The Smurfs* (orig. *Les Schtroumpfs*) is very interesting in this respect, as it involves all these forms (with thirty-nine comic albums created between 1958 and 2022,[13] various animated movies and also television series). All these texts host similar representations of Mother Nature, who is depicted as an elderly figure with power over nature patterns living in the Smurfs Forest. Correspondingly, she has white (or rarely blonde, in the comic books) hair tied up in a bun by a flower garland, wears a long pink bell-bottomed dress over her rounded body (which recalls the idea of abundance found in ancient representations), and always carries a star-topped wand with her (see Fig. 3.5).

However, her magic is rather limited: in fact, she mostly simply ensures that natural rhythms (e.g. the succession of seasons) develop properly (e.g. by turning bushes and trees red in autumn,[14] or making them flourish in springtime[15]), and is herself shown

[12] I.e. an effect of "expulsion" from the domain of enunciation (cf. Greimas and Courtés 1979; Engl. Trans. 1982: s.v. "disengagement").

[13] Sixteen by its inventor, the Belgian comics artist Peyo, the others by his studio. For an interesting analysis, see in particular Eco (1983 [1979]).

[14] Such as in the episode "All Hallows' Eve" of season 3 from the Smurfs cartoon show, first aired on May 11, 1983, where Mother Nature, after having mistakenly turned Lazy Smurf red, apologises for this and says that she cannot change him back to blue with her red wand, nor does she own any blue wand, as she only possesses the colours needed for the different seasons. She then leaves him in tears as she says: "But I can't do anything now, I have work to do; autumn cannot wait further time like this just so" (8' 48″–8' 59″) (see Fig. 3.5, upper images).

[15] Such as in the episode "The Smurfs Springtime Special" of season 1 from the Smurfs cartoon show, first aired on April 8, 1982, where Mother Nature awakens to get ready for the first day of spring (see Fig. 3.5, lower images).

**Fig. 3.5** Screenshots from the episode "All Hallows' Eve", 1983 (top) and "The Smurfs Springtime (Special)", 1982 (bottom) from *The Smurfs* cartoon show[16]

as partially dependent on them (such as when she laments the weakening effects of winter on her "old bones"[17]).

Moreover, she does not seem immune from others' magic spells, as shown in the episode "The Smurfs Springtime (Special)" (1982), when Gargamel (i.e. the arch-nemesis of the Smurfs, a man—!—and magician who plots against them) gets her to smell a bouquet of magic flowers before she completes her spring duties, thus making her faint and fall under his power, and stopping spring itself. It is also interesting to note that in the 2021 series Mother Nature is said to have retired,[18] and is replaced by a new fairy called Leaf, which further highlights her limited super-natural characterisation.

A more powerful figure is featured in the 1993 American animated musical fantasy film *Happily Ever After* (originally released as *Snow White: The Adventure Continues* in the Philippines in 1989), written by Robby London and Martha Moran, and directed by John Howley. Here Mother Nature is depicted as the most powerful force of

[16] Videos posted by "Don Israel" on Fandom, https://www.fandom.com (specific links available in the references).

[17] See the same episode dedicated to springtime mentioned above.

[18] By contrast, as Eco (1983 [1979]: 265) ironically remarked, the Smurfs seem not to be subjected to time, with the only exception of Papa Smurf: "Dunque i puffi vivono nella foresta, sono blu, piccolissimi, di età indefinita, salvo il Gran Puffo, che è vecchio e ha la barba bianca (i puffi vivono in una società gerontocratica perfetta dove tutti sono più o meno infanti e c'è solo un anziano, depositario autoritario ma paterno di tutta la saggezza, …)" [*So the Smurfs live in the forest, they are blue, very small, of indefinite age, except for Papa Smurf, who is old and has a white beard (the Smurfs live in a perfect gerontocratic society where everyone is more or less an infant and there is only an elderly, authoritarian but paternal custodian of all wisdom …)*, my translation].

good, having complete control over nature, as well as the ability to create creatures from potions she makes in her "natural laboratory". However, her creations are not impeccable (as a sequence of the movie ironically highlights, playing on the visual representation of animals whose names are compound words, such as the "catfish", the "bullfrog", etc.), and she seems to proceed according to a "regime of adjustment" (in the terms introduced by Landowski 2005) rather than by a real "programming" or "manipulation" strategy (cf. *Ibidem*).[19] She is personified by a middle-age, rather eccentric woman, with red hair and lipstick, an unusual hat made of pink leaves and yellow flowers, a yellow dress and a green furry boa decorated with purple flowers wrapping her shoulder and waist and highlighting her curvy hip (see Fig. 3.6).

**Fig. 3.6** Screenshots from the trailer of *Happily Ever After* (Howley 1993)

Mother Nature is also a recurring character in *The New Woody Woodpecker Show*, an animated television series based on the original cartoon *The Woody Woodpecker Show* (Smith et al. 1957–1972) by Walter Lantz, produced by Universal Animation Studios and aired from 1999 until 2002 on FOX (United States) and BBC (United Kingdom), with characters from the classic series (e.g. Woody Woodpecker, the main character) and a few new ones (among which Mother Nature herself).

[19] According to the author of *Les interations risquées* (Landowski 2005), four different regimes underlie meaning-making processes: the "regime of programming" (*régime de la programmation*) is based on a principle of regularity and arises when the aims previously set by the involved subject are achieved; the "regime of manipulation" (*régime de la manipulation*) is founded on the logic of intentionality, and follows the classic model of interaction between a subject and an object; the "regime of adjustment" (*régime de l'ajustement*) relies on perception and the progressive acquisition of special skills, therefore expressing insecurity; finally, the "regime of accident" (*régime de l'accident*) is based on the logic of chance and risk, and opposes the system of programming as it is untied from any pre-established behaviour.

**Fig. 3.7** Screenshots from *The New Woody Woodpecker Show*[20]

In this case, she is depicted as a fairy with red or white hair (depending on the episode) and butterfly wings, wearing a pink (or red, depending on the episode) dress with a white apron, a gold crown, red lipstick, and a wand with a golden star (see Fig. 3.7, upper part). Sometimes she also has a clipboard (Fig. 3.7, third image), since her main duty is to check and ensure that natural events develop properly. For this reason, she is often shown while making sure that Woody does his part in nature, also being authoritative when he refuses to do so. In the first segment of the 14th episode of season 2, for instance, when she finds him relaxing instead of taking care of his duties (see Fig. 3.7, last image), she reproaches and punishes him by forcingly sending him to Woodpecker school, where he meets Woodrow Woodpecker and compete with him in a series of tasks.

As a protector of the natural order, the figure of Mother Nature also plays a crucial role in the animated movie *Epic* (Wedge 2013), where she is personified by Queen Tara,[21] the ruler of the secret world of Moonhaven. However, the vulnerability of this figure is also particularly stressed: while tenacious and determined to defend her kingdom and its inhabitants at any cost, Tara is at the same time one of the most delicate components of nature—and hence has an entire army protecting her.

[20] The first image was posted by Roger Blake on Fandom, https://walterlantz.fandom.com/wiki/Mother_Nature?file=Woody_Woodpecker_-_Downsized_Woody_-_02_-_Mother_Nature.jpg; the other three images were posted by Don Israel on Fandom, https://walterlantz.fandom.com/wiki/Teacher%27s_Pet (Accessed: 12 December 2022).

[21] A denomination that recalls the Roman *Terra* or *Terra Mater*.

**Fig. 3.8** Screenshots from *Epic*[23]

On the figurative level, a floral characterisation is used, in this case too, to highlight her relation to the natural world, opening the way to a real "fusion": the Queen's body is not only enclosed by, but to some extent also merged with a sort of "inverted" bud, with a leafy upper part covering her chest and a white, petalous bell-shaped skirt wrapping her legs—as if they were the pistils of a flower (see Fig. 3.8). What is more, she seems to be directly connected to the natural dimension, having the power to control roots, branches and leaves with her mind and actions.[22] Her magic also includes the ability to heal and regrow killed plants, and allows her spirit to remain in the royal pond after her death, until she chooses her heir (a marigold jinn whom she had rescued earlier that day).

Sometimes, Mother Nature's magic powers are more openly combined with (or rather opposed to) others'. This happens, for instance, in the special *The Year Without a Santa Claus* (Rankin and Bass 1974), where she is presented as the mother of the conflicting Miser Brothers: Snow Miser, who controls cold weather, and Heat Miser, who controls warm weather. Represented as an aged curvy woman (Fig. 3.9, left), Mother Nature wears in this case a green dress with floral decorations and a green tulle hat that functions as a real nest for birds, thus stressing her relationship with the environment and her role in the natural order. She administers her duties from an isolated location among the clouds where she serves tea to visitors and keeps various forest animals as company. Her magic powers include physical displacement

[22] In the scene from which the frames represented in Fig. 3.8 are taken, for instance, the leaves and lily pads around her move towards her feet, allowing her to walk on water.

[23] Posted by ChuckandSarahmyotp on Fandom, https://epic-the-movie.fandom.com/wiki/Queen_Tara?file=Epic-movie-screencaps.com-2724.jpg (first image) and https://epic-the-movie.fandom.com/wiki/Queen_Tara?file=Epic-movie-screencaps.com-2742.jpg (second image) (Accessed: 12 December 2022).

**Fig. 3.9** Screenshots from *The Year Without a Santa Claus* (left) and *A Miser Brothers' Christmas* (right)[24]

transportation by way of lightning, which, together with her authoritative and rather severe personality, makes most characters perceive her as intimidating.

Mother Nature appears also in the sequel of the special, *A Miser Brothers' Christmas* (Barton Thomas 2008), where she is revealed to be the mother of more children, each representing a different natural phenomenon: Earthquake, Thunder and Lightning, the Tides and the Winds (also including the North Wind, who plays the role of the villain and is punished by Mother Nature herself in the end of the special). While her new design makes her look younger and more stylish (Fig. 3.9, right), her characterisation remains mostly unaltered—also on the visual level, with continuity ensured by her green dress, her thick head of green hair (chromatically recalling the hat worn by her 1974 antecedent), which now hosts a bunch of flowers instead of a bird nest, and the presence of flowers on both her jacket and skirt.

Mother Nature's links with other magic characters are also emphasised in two titles in *The Santa Clause* trilogy, *The Santa Clause 2: The Mrs. Clause* (Lembeck 2002) and *The Santa Clause 3: The Escape Clause* (Lembeck 2006). Head of the Council of Legendary Figures (which also consists of Santa Claus, Easter Bunny, Cupid, Father Time, Sandman, Tooth Fairy and Jack Frost), she is first seen at one meeting, held at the North Pole, and then serving as the officiate at Santa's wedding to Carol Newman.

Her leading role is figurativised by the majestic gold crown over her head (see Fig. 3.10) and further confirmed in *The Santa Clause 3*, where she proposes a disciplinary action against Jack Frost as a reaction to his attempt to upstage Santa Claus and his violation of the Legendary Figures code of conduct. However, her power seems mostly limited to the official domain (not without motions and counteractions by the other members of the Council, sometimes with effects on the results of her

[24] Posted by PyroGothNerd on Fandom, https://christmas-specials.fandom.com/wiki/Mother_Nature?file=Heat-miser-snow-miser-valiantfans-com.jpg (first image) and https://christmas-specials.fandom.com/wiki/Mother_Nature?file=The_punishment_they_get_by_caressechris-d4gnht8.jpg (second image) (Accessed: 12 December 2022).

**Fig. 3.10** Screenshot from *The Santa Clause 2: The Mrs. Clause*[25]

decisions), as demonstrated, for instance, by her inability to affect the magic of other Legendary Figures (which further aligns her to them).

In addition to the ornament covering her head, other visual features characterising this figure should be noted. More specifically, her dress marks an evident difference from the previously analysed cases, replacing the use of green with a less connoted purple, and the presence or representation of flowers with the inclusion of a number of thin golden filaments, which seem to evoke the form of natural roots, but could also simply function as elegant adornments. A particular mention to the ethnical characterisation of this Mother Nature should also be made, which confirms a new trend in her representation, already identifiable in the figure of Queen Tara in the movie *Epic*.

Physical variety is also sought in another case, the Italian variety game show *Ciao Darwin*[26] (Recchia 1998–2003; Cenci 2007–2019). Since the first episode, this TV programme has hosted various personifications of Mother Nature, with a single representative (generally a model and/or an actress, in any case a beautiful and statuesque young woman) in the first two editions (1998 and 1999), and almost thirty different women of various origin and physical appearance in the following episodes (seasons 3–8, aired from 2000 to 2019). Since season 3 (2000), sporadic[27] appearances of "Father Nature" (himself personalised by different attractive male

[25] Posted by WMSWMCQSADAKJM on Fandom, https://disney.fandom.com/wiki/Mother_Nature?file=Mothernature.jpg (Accessed: 12 December 2022).

[26] Whose format has been sold in various countries of the world, including Greece, Hungary, Poland, Romania, Serbia, China, and Vietnam. I focus here on the original Italian version.

[27] Five in total.

models and/or actors) were also introduced, thus extending diversity aspects to the gender dimension. Nevertheless, it must be noted that a process of de-subjectification seems to be at work in this case, reducing all these figures to mere "objects of vision"(according to the terminology introduced by Berger 1972[28]). In fact, while their main role consists in spinning a globe to randomly select the competitors who have to participate in various games, the moment when they generally receive most attention is when they make their entrance. Despite the peculiarities of each case, a recurring scheme can be identified in the development of this part of the show: after the presenter (Paolo Bonolis), sometimes accompanied by his co-host (Luca Laurenti), introduces her (or him) by highlighting their role in the creation and maintenance of the natural order[29]—to which evolution, which is the main idea underlying the TV programme, is reconducted[30] —, Mother (or Father) Nature enters the TV studio (coming from a door decorated with drawings of rocks, a waterfall and wild vegetation in the first editions, or descending a high, central stairway, with a billowing smoke effect coming upward from the floor and a video of a waterfall projected just behind her/him, in the most recent seasons), while the song *Adiemus*, based on a mix of melodies in African and Celtic tribal style,[31] comes to dominate the soundtrack. All with a distant look and a serious expression on their face, the "elected" women and men walk slowly and confidently, while the camera zooms on their refined bodies, which are anything but "natural" (as they are evidently modelled by hard exercises in the gym, strict diet regimens, frequent sessions at the beautician, when not even surgical interventions), and are almost totally exposed, with only

[28] The reference is here to John Berger's famous reflections reported in *Ways of Seeing*: "A woman must continually watch herself. She is almost continually accompanied by her own image of herself. … Her own sense of being in herself is supplanted by a sense of being appreciated as herself by another. … One might simplify this by saying: *men act* and *women appear*. Men look at women. Women watch themselves being looked at. This determines not only most relations between men and women but also the relation of women to themselves. The surveyor of woman in herself is male: the surveyed female. Thus she turns herself into an object—and most particularly an object of vision: a sight" (Berger 1972: 46–47). The same dynamics seem in this case to concern the figure of Father Nature too.

[29] E.g. by recurring to expressions such as "colei che tutto puote" [*the one who can do everything*], "colei che tutto muove" [*the one who moves everything*], "colei che decide la gradazione di tutti gli ingredienti" [*the one who decides the gradation of all the ingredients*], "Lei, che determina le sorti di tutti noi" [*She, who determines the fate of all of us*]. Similar expressions are also used to introduce Father Nature—e.g. "L'incanto primigenio che tutto muove. Voi, donne, preparatevi all'incanto" [*The primordial enchantment that moves everything… Women, get ready for the enchantment*, all translations are mine].

[30] The sentence used in the first episode is particularly interesting in this respect: "Giacché la natura è la dominante di ogni evoluzione, la natura è proprio qui. [Signore e signori,] Madre Natura" [*Since Nature presides over all evolution processes, Nature is right here. [Ladies and gentlemen,] Mother Nature*, my translation].

[31] As its composer Karl Jenkins observed, the song was "written phonetically, with the words viewed as instrumental sound" in the aim to "create a sound that [wa]s universal and timeless" (in Barone 2016), without transmitting any specific message (see Taylor 2007). In this sense, it can be considered a case of "musical glossolalia" (as described in Stano 2019).

minimal clothing covering their most intimate parts.[32] Mother (or Father) Nature, in other words, is here almost completely "unveiled", that is, "laid bare" by a voyeuristic gaze that looks at her (or his) body as pure, objectified "flesh", rather than a sensitive and subjective entity.

While such an exposition takes place, the host celebrates the "beauty of nature", by means of verbal appraisals ("la bellezza della natura" [*the beauty of nature*], "è la natura, è la natura, è la natura... è la bellezza della natura" [*It is nature, it is nature, it is nature... it is the beauty of nature*, my translation]), as well as through paraverbal signs expressing amazement and astonishment, which are generally echoed (and further emphasised) by the men and women in the public—who simulate fake fainting, show off real or simulated binoculars, or bring their hands to their heads to express their astonishment for such beauty, while appreciative chants and whistles (according to the most stereotyped canons of *machismo*, which here hyperbolically become non-gendered ways of expression) overlap each other.

It should also be noticed that such a voyeuristic and objectifying process does not stop here. Mother/Father Nature stands still or seated for almost the entire duration of the show, keeping smiling at the camera in absolute silence (or at most simply exchanging a few words with the hosts), while sketches of various kinds take place around her/him, sometimes also making fun of their inability to understand what is happening. Even when they are more actively involved, such as in the "underwear show", it is basically to insist on their appearance, never getting to their identity (or "essence").

What is more, there is no shortage of sexual—and often also sexist—jokes or sketches, as exemplified by the very first episode, when Laurenti comments on the fact that, being the woman who just entered the television studio "Mother Nature", he and Bonolis are to be considered her sons, which—as he remarks ironically, alluding to the impossibility of having sex with her—"is a real shame" ("peccato!").

All these elements confirm an extremely simplistic, trivialised and sexualised vision of Mother Nature, who does not *act*, but rather *appears* (to recall Berger's reflections), losing any divine characterisation and power to be reduced to a mere object of sight, whose "own sense of being in herself is supplanted by a sense of being appreciated as herself by another" (Berger 1972: 46).

[32] In fact, in the first season, Mother Nature's body was covered by a long tunic, which, however, being semi-transparent, allowed her body, including intimate parts such as her nipples and glutei, to be half-seen. Moreover, from season 2, the tunic was abandoned, opting for minimal loincloths and bras, generally flesh-coloured and with a few artificial flowers sewn on them. The same applies to Father Nature, whose mostly naked body and generally long hair clearly evoke the typical representation of Tarzan, the archetypical "wild man". Irony has also sometimes been used to stress the importance of physical appearance: in the 13th episode of season 5 (aired on 15 December 2007), for instance, a fat rather small man is presented as Father Nature, and "greeted" with hilarity and boos of disapproval; a similar reaction was registered when Laurenti himself showed up disguised as Mother Nature in the 5th episode of season 8.

### 3.3.3 Mother Nature in Advertising

Mother Nature has been a recurrent character also in advertising, resulting in different figures and effects of meaning.

In a 1970s commercial for Chiffon Margarine, for instance, she is personified by a middle-aged woman (actress Dena Dietrich), who appears on the screen in a long white dress, simply adorned with white flowers, just like her hair. She appears on the screen while she enters a jungle-like scenario by swinging on a vine and making a Tarzan-like yell ("Oohoohohoohooohooh"). Directly addressing her, the off-screen narrator then informs her that she has mistaken Chiffon margarine for butter, and she responds, directly looking back at the camera, with the trademarked slogan "It's not nice to fool Mother Nature", underlined by a thunder and the subsequent running of an elephant (whose trumpeting closes the video) towards the camera itself. The emphasis is therefore put in this case on wilderness, but also on a conception of nature as the realm of the "ordinary", or established conventions (i.e. the result of a process of "naturalisation"), which makes the extra-ordinary features of the sponsored product "fool Mother Nature", with a marked ironic effect.

Irony is also a key component of the international campaign "Outsmart Mother Nature" (2007–2009) by Leo Burnett for Tampax feminine hygiene products. In this case, Mother Nature is represented as a brash, intrusive middle-aged woman in a green tweed skirt-suit handing a "monthly [red] gift" (symbolising menstruation) to young women. In fact, as one of the executive directors of the campaign, Anna Meneguzzo, declared in an interview, "Mother Nature was the perfect enemy. Nature behaves sometimes as a stepmother rather than a mother" (2018, in Røstvik 2020: 429). She was therefore designed to be "a villain of sorts", a sort of "mother-in-law" (*Ibidem*), drawing on crucial figures in the collective imaginary, such as the fashionable, unfriendly and powerful boss Miranda Priestly from *The Devil Wears Prada* (Frankel 2006) or Bree Van de Kamp from the TV series *Desperate Housewives* (2004–2012), "a neurotic and perfectionist homemaker in turn inspired by the Stepford-wife archetype first conceptualized in Ira Levin's 1972 satirical thriller about robotic wives" (Røstvik 2020: 437). As a counterpart to this figure, the campaign hosted various young independent women receiving Mother Nature's visit (and "gift") while in a high-risk situation for menstruation (e.g. dancing,[33] posing for a photo shooting,[34] travelling to exotic places in short white dresses,[35] sleeping,[36] dreaming about romantic situations with a man,[37] and so on). Initially surprised and apparently disturbed by her arrival[38] and her request to interrupt what they are doing,

---

[33] https://www.youtube.com/watch?v=kOJeYbpyDSM (Accessed: 15 November 2022).

[34] https://www.youtube.com/watch?v=9kDT3xmj9bU (Accessed: 15 November 2022).

[35] https://www.youtube.com/watch?v=HE85CEM_0zY and https://www.youtube.com/watch?v=J3kY50vV-rQ (Accessed: 15 November 2022).

[36] See, for instance, the example shown in Røstvik (2020: 430).

[37] https://www.youtube.com/watch?v=O5ZQdDwncCc (Accessed: 15 November 2022).

[38] In the commercials, for instance, they all say: "Mother Nature, not now!".

such women promptly "outsmart" her, overturning her plans thanks to the "empowering" product promoted by the campaign. The idea was later further enhanced by involving the famous tennis player Serena Williams[39] in new posters[40] and YouTube videos,[41] thus reinforcing the veridictory contract of the brand—as she represented, in fact, a powerful young woman, effectively able to "outsmart" Mother Nature and her "gift" in her professional life. Thus, the figure of a maternal, benevolent goddess was evidently overturned, interpreting Mother Nature as an opponent (in Greimassian terms), that is to say, as an enemy to be defeated—and which can actually be defeated through "culture" (i.e. the science standing behind the sponsored product, whose results are particularly stressed in the end of commercials, as well as in the notes included in the print advertisements).

The dichotomy nature/culture also characterises another campaign, the one realised for Glaceau's VitaminWater10 in 2009. Here Mother Nature is represented as a powerful and rather bossy businesswoman (i.e. the CEO of a company named Water Incorporated) who directs a team of "wild"—in the literal sense of "[usually] living or growing in the natural environment" (Oxford Dictionary 2022: s.v. "wild"), but also "uncontrolled"—animals.

In one commercial,[42] for instance, she informs her staff—composed of three prairie dogs on a desk, with a wolf sitting just behind them, a squirrel and a deer—that they have to work late, as "VitaminWater10 just hit the market. It's 10 cal per serving, it's packed with vitamins, and it's *naturally* sweetened" (0′ 05″–0′ 11″). Then she provocatively adds: "Natural? Come on. I'm Mother Nature. We need to take back our competitive edge. We will work all night if we have to" (0′ 12″–0′ 19″), thus anticipating the message explicated by the final claim: "It's one-upped nature" (0′ 29″). Thus the advertisement seems to suggest the superiority of culture over nature, as it is also ironically highlighted by the behaviour of Mother Nature's employees. In fact, during her speech, when the camera is directed to her, a gulp is heard. Then the commercial cuts back to the desk hosting the three prairie dogs, and one of them is missing, with only some fluff left in its place. Subsequently, another gulp is heard while the camera moves back to the woman, and as it goes back to the desk, another prairie dog is missing. The third prairie dog then asks if they could "order some take-out" before going back to work, making Mother Nature realise what happened: because of its "wilderness"—i.e. nature—the wolf could not resist its primordial instinct to eat the prairie dogs, even if they were his co-workers. The unreliability of the natural team is further confirmed in other advertisements,[43] which ironically insist on the wild animals' inability to control sexual instincts or silly behaviours, despite Mother Nature's warnings and admonitions. On the contrary,

---

[39] Directly confronting—and always beating—Mother Nature in tennis matches, public interviews, etc.

[40] See, for instance, the examples shown in Røstvik (2020: 432–433).

[41] E.g. https://www.youtube.com/watch?v=HIeYV63ap20 (Accessed: 15 November 2022).

[42] https://www.youtube.com/watch?v=u9nQ-0njLHo&t=30s (Accessed: 15 November 2022).

[43] See, for instance, https://www.youtube.com/watch?v=tFMg8ldDmHQ and https://www.youtube.com/watch?v=iQh1tILW7O4 (Accessed: 15 November 2022).

culture is shown as able to "cut into nature", without having to deal with these kinds of problems.

Occasionally, while always stressing the opposition between nature and culture, marketing communication rather focuses on the negative effects of the latter onto the former. This is the case, for instance, of the 2014 campaign "Nature is speaking"[44] promoted by the American non-profit environmental organisation Conservation International, which includes a video specifically devoted to Mother Nature.[45] No humanlike personification takes place here, and a relation is rather established with various images of wilderness (e.g. mountain peaks in the clouds, waterfalls, moving seaweeds, growing mushrooms, forests, snowy mountains, disintegrating glaciers, desert dunes, sea waves, raging rivers, canyons, icebergs, … and finally a partial view of the Earth from space), while a voice-over says:

> Some call me 'nature'. Others call me 'Mother Nature'. I've been here for over 4.5 billion years, 22,500 times longer than you. I don't really need people. But people need me. Yes, your future depends on me. When I thrive, you thrive. When I falter, you falter. Or worse. But I've been here for eons. I have fed species greater than you, and I have starved species greater than you. My oceans. My soil. My flowing streams. My forests. They all can take you, or leave you. How you choose to live each day, whether you regard or disregard me, doesn't really matter to me. One way, or the other. Your actions will determine your fate, not mine. I am Nature, I will go on. I am prepared to evolve. Are you?

As the title chosen for the campaign suggests, in this case "Nature [itself] is speaking". In fact, through a reiterated *embrayage*, Mother Nature confirms the identification suggested by the visual dimension (e.g. "*my* oceans", "*my* soul", "*my* flowing streams", "*my* forests"), also warning her addressees about the risks of imposing culture ("*your* actions") over her. Though always connected with the female universe (through the choice of the voice-over, which is actress Julia Roberts'), Nature is evidently detached from any anthropomorphic—or "cultural"—characterisation, and actually abandons her maternal attitude towards humanity, to take on a rather threatening tone. A sharp opposition between nature and culture is thus reaffirmed (as the use of the personal pronouns and adjectives clearly show), but this time to support the superiority of nature over culture (as the catastrophic scenarios mentioned by the voice-over suggest and the final claim further confirms, stating that "Nature doesn't need people. People need Nature").

A more optimistic, though always controversial, view characterises Posten[46]'s recent Christmas advertisement (2022),[47] which figurativises the tension between nature and culture through a sentimental relationship between Mother Earth (representing the former) and Father Christmas (symbolising the latter). "In the beginning"—Father Christmas' voice-over says, reporting what he is writing in a letter to Mother Earth—"everything was fine. We were made for each other. Polar opposite, but still a perfect match" (0′ 04″–0′ 19″). While the characters appear on the screen

[44] https://www.conservation.org/nature-is-speaking (Accessed: 15 November 2022).

[45] https://www.youtube.com/watch?v=WmVLcj-XKnM&t=2s (Accessed: 15 November 2022).

[46] Postal and freight services in Norway and abroad.

[47] https://www.youtube.com/watch?v=cRM5PpsAH4g (Accessed: 16 December 2022).

holding their hands and showing affection to each other, the commercial interestingly alludes to a possibility of encounter between the two different "poles" they represent, though only apparent: "We had a shared purpose, to give children what they wanted. Maybe we were naïve? I miss being naïve. I miss that you were happy and that we didn't fight" (0′ 22″–0′ 56″). In fact, Mother Earth soon becomes sick (both physically and emotionally) and starts reproaching Father Christmas for his behaviour. A flashback is therefore introduced to better described their fights and separation:

> Mother Earth [after receiving a bag full of unspecified objects addressed to Father Christmas]: "How long do you think we can go on like this?"
>
> Father Christmas: "What do you mean?"
>
> Mother Earth: "You don't see me anymore! When is it enough? It's always about you. More, more, more!"
>
> Father Christmas: "What's wrong with making others happy?"
>
> Mother Earth [in an evidently upset tone]: "You don't think about the consequences! You don't take any responsibility!"
>
> Father Christmas [while pointing to an object in his own hands]: "What you dislike about me that's what everybody else loves about me."
>
> Mother Earth [shouting]: "But don't you see that it's too much? What world are you living in?"
>
> [Father Christmas breaks a vase while he tries to get closer to Mother Earth]
>
> Father Christmas [annoyed]: "Now we have to by a new one!"
>
> Mother Earth [in tears]: "This is all we've got!"
>
> Voice-over [flashforward]: "I admit that I didn't see how serious it was. I just didn't want to acknowledge that things weren't working out between us."
>
> Father Christmas [in tears, while Mother Earth approaches the door, ready to leave]: "You know that I do it for the children, right?"
>
> Mother Earth [in tears, just before leaving]: "So do I".
>
> (Posten 2022, 1' 30″–2' 34″).

A new flash-forward brings the spectator back to Father Christmas, alone and unhappy in his house, regretting his behaviour, and then suddenly changing expression as he looks outside of the window. After receiving his letter (which, as the voice-over clarifies, he concluded precisely by recognising that he "can't live without [her]. No one can", 3′ 20″–3′ 23″), Mother Earth decided to come back (i.e. to give him, i.e. man, or "culture", another chance). However, a final warning appears on the screen to mark that this is not precisely a happy ending: "Mother Earth has no more chances to give. In 2022 only 10 of the largest companies in Norway cut their emissions in line with the Paris Agreement. **We must all step up.** From: **Words** To: **Actions**" (3′ 36″–3′ 50″).

The commercial therefore reiterates the idea of a sharp opposition between nature, represented by an austere unadorned woman, and culture, symbolised by the consumeristic culture generally associated with Santa Claus in current societies. While initially attenuated, such an opposition is shown as increasingly growing, and being destined to break the relation between these "polar opposites", making Nature leave Culture, unless the latter recognises its faults and follows the former's needs and requirements. Nature's motherhood, in other terms, is put at risk by culture itself, as nature "has no more chances to give"—which leads, significantly, to the final *embrayage*, used as a way to call the spectators directly into action: "We must all step up".

## 3.4 Concluding Remarks

Although certainly not exhaustive of all the existing representations of Mother Nature, the discussion provided above reveals a series of relevant recurring patterns in its figurativisation and signification, opening the way to interesting reflections on the Western contemporary collective imaginary, also in comparison with its traditional iconography.

First of all, a process of "secularisation" can be observed, which echoes and further enhances the dynamics initiated in the Renaissance and the modern era. In fact, the considered case studies seem to confirm a process of "blurring" of the traditional idea of "Mother Nature" as a sacred generative principle, making her lose her divine connotation and aligning her to other legendary figures, such as fairies and other mythical characters in Western folklore (e.g. Santa Claus, Easter Bunny, Sandman, Cupid, etc.).[48] Consequently, her powers and roles have evidently changed: no longer a creator of the world and its inhabitants, nor a nurturing source,[49] Mother Nature is rather conceived as a simpler "supervisor" of the natural order, which ensures its respect and proper development (as it happens, for instance, in *The New Woody Woodpecker Show*, as well as in *Epic*, *The Year Without a Santa Claus*, *A Miser Brothers' Christmas* and *The Santa Clause* trilogy), or at most functions as its "executor" (as exemplified by the case of *The Smurfs*, where Mother Nature's duties mainly consist in making the flowers bloom in spring and turning the leaves red in autumn). Only in a few cases a generating impulse is still present, but is nonetheless presented as defective, as evidenced by Mother Nature's peculiar creations and her

[48] Which seems to have made her become a typical character of Christmas specials and other texts primarily addressed to children.

[49] In fact, the absence of the emphasis traditionally put on feeding (e.g. by representing several breasts) is evident. On the contrary, the texts related to the foodsphere (e.g. Chiffon and Vitamin-Water10 commercials) ironically insist precisely on a supposed superiority of the cultural processes of production of food. This, however, evidently contrasts with a widespread trend in contemporary cultures, where a particular praise for "natural foods" exists—as Chap. 5 will illustrate in further detail.

subsequent necessary adjustments in *Happily Ever After*, as well as by her inability to contain her assistants' "wilderness" in the commercials promoting VitaminWater10.

Accordingly, her personification has also evidently changed: apart from a very few representations highlighting extra-ordinary features, such as the fairy wings that Mother Nature has in *The Woody Woodpecker Show* or the inhabited bird nest hosted on her head in *The Year Without a Santa Claus*, she is mainly reduced to the limited "corposphere" (cf. Finol 2021) of common women (at most with particular suits or ornaments highlighting her leadership, such as in the advertisements for Tampax and VitaminWater10 and the *Santa Clause* trilogy, or a particular chromatic configuration or adornments evoking her relation to nature, such as in the movies *The Year Without a Santa Claus* and *A Miser Brothers' Christmas*, or the Chiffon margarine commercial). What is more, in some cases she is subjected to a real process of de-subjectivation, which makes her lose any "essence", to become the object of a voyeuristic vision[50]—as it can be observed, for instance, in the case of the TV show *Ciao Darwin*, as well as in some verses of McLean's song *Mother Nature*. Finally, it must be noted that sometimes (e.g. in the video *Nature is speaking*, as well as in the song *Death of Mother Nature Suite*) Mother Nature loses all anthropomorphic characterisations, as a way to stress the separation (and opposition) between nature and culture. These are generally the cases when she is presented as dead or dying, precisely as a result of the action of man (i.e. the culture to which it is opposed) and its "catastrophic"[51] consequences.

Therefore, a marked dysphorisation of Mother Nature emerges, as she ceases to be a figure who generates, nourishes and protects humanity, to take on a more demanding—as exemplified by the reproaching figure appearing in *The New Woody Woodpecker Show*, or the intrusiveness and authoritarianism projected on the characters appearing in the ads for Tampax and VitaminWater10 —, intimidating—as evidenced in particular in the movies *The Year Without a Santa Claus* and *A Miser Brothers' Christmas* —, and self-centred attitude—as highlighted by both the campaign promoted by Conservation International (which insists on the fact that "Nature doesn't need people. People need Nature") and Posten's commercial (which figurativises this same aspect by insisting on Mother Earth and Father Christmas' breakup). A counterpart to such a dramatizing trend, which insists on the loss or estrangement of Mother Nature, is its "ironicisation", which plays with and on the figure of Mother Nature, in a sarcastic (e.g. the ads for Tampax, VitaminWater10 and Chiffon margarine) and sometimes even mischievous (e.g. *Ciao Darwin*) way. In any case, nature seems unable to escape an eminently dualistic perspective, which strongly opposes it to culture, and presents any attempt to combine these two dimensions as "naïve" (such as in the commercial by Posten) or inevitably destined to fail—either by means of a progressive separation (as it happens with Mother Earth

[50] As a result, as highlighted above, the body of Mother Nature is almost totally exposed, and abandons the rounded forms of abundance and fertility to adhere to contemporary standard of beauty and attractiveness. In fact, nakedness and corporeality are very interesting as related to the conception and representation of nature, as I will discuss in detail in Chap. 6.

[51] The figure of the "catastrophe" plays in fact a crucial role in contemporary discourses on nature, as discussed in detail in Chap. 4.

and Father Christmas in the same advertisement) or overcoming (as in the ironic advertisements by Leo Burnett and Glaceau), and sometimes even a real annihilation (i.e. when the emphasis is put on the Mother Nature's death, but also when she becomes the sexualised object of a voyeuristic look).

## References

Barone, Brian. 2016. An aural history of "adiemus". *The Awl*, Sept 16. https://www.theawl.com/2016/09/an-aural-history-of-adiemus/. Accessed 13 Dec 2022.

Berger, John. 1972. *Ways of Seeing*. London: BBC and Penguin Books.

Blasius, Gerardus Leonardus. 1681. *Anatome animalium Terrestrium Variorum, Volatilium, Aquatilium, Serpentum Insectorum, Ovorumque, Structuram Naturalem*. Amsterdam: Joanis á Someren/Hernrici & Viduæ Theodori Boom.

Brevini, Franco. 2013. *L'invenzione della natura selvaggia*. Torino: Bollati Boringhieri.

Collignon, Maxime. 1890. *Manual of Mythology, in Relation to Greek Art*. London: Grevel & Co.

Dransart, Penny. 1992. Pachamama: The inka earth mother of the long sweeping garment. In *Dress and Gender: Making and Meaning*, ed. Ruth Barnes and Joanne B. Eicher, 145–163. New York, NY, and Oxford, UK: Berg.

Eco, Umberto. 1983 [1979]. Schtroumpf und Drang. Schtroumpf und Drang. *Alfabeta*, Sept 5. Republished in Id. 1983. *Sette anni di desiderio. Cronache 1977–1983*, 265–271. Milan: Bompiani.

Finol, José Enrique. 2021. *On the Corposphere. Anthroposemiotics of the Body*. Berlin: De Gruyter.

Gravelot, Hubert von e Charles Nicolas Cochin. 1791. *Iconologie par figures ou Traité complet des Allégories*. Paris: Lattré [engraver s.d.].

Greimas, Algirdas J., and Joseph Courtés. 1979. *Sémiotique: Dictionnaire raisonné de la théorie du langage*. Paris: Hachette. English Trans. Crist, Larry, Daniel Patte, James Lee, Edward McMahon II, Gary Phillips, and Michael Rengstorf. 1982. *Semiotics and Language: An Analytical Language*. Bloomington, IN: Indiana University Press.

Hadot, Pierre. 2004. *Le voile d'Isis : essai sur l'histoire de l'idée de nature*. Paris, Gallimard. English Trans. Chase, Michael. 2006. *The Veil of Isis: An Essay on the History of the Idea of Nature*. Cambridge, MA, and London, UK: The Belknap Press of Harvard University Press.

Harding, Elizabeth U. 1993. *Kali: The Black Goddess of Dakshineswar*. York Beach, ME: Nicolas-Hays.

Izzo, Joshua. 2022. *The World of Avatar: A Visual Exploration*. London: DK.

Kempt, Wolfgang. 1973. *Natura. Ikonographische Studien zur Geschichte und Verbreitung einer Allegorie*. Dissertation, Tübingen.

Landowski, Eric. 2005. *Les interactions risquées, Nouveaux Actes Sémiotiques* 101–103. Limoges: Pulim.

Leopardi, Giacomo. 1835 [1824]. "Dialogo della Natura e di un Islandese". In *Operette Morali*. Napoli: Saverio Starita. English Trans. Edwardes, Charles. 1882. "Dialogue Between Nature and an Icelander". In *Essays and Dialogues of Giacomo Leopardi*, 73–79. London: Trübner & Co.

Liddell, Henry George, and Robert Scott. 1940. *A Greek-English Lexicon*, revised and augmented by Sir Henry Stuart Jones, with the assistance of Roderick McKenzie. Oxford: Clarendon Press.

Long, George, ed. and trans. 1880 [1862]. *The Thoughts of the Emperor M. Aurelius Antonius*, 2nd ed. revised and corrected. London: George Bell & Sons.

Macrobius (Ambrosius Theodosius). 2011 [403]. *Saturnalia*, ed. and trans. Robert A. Kaster. Cambridge, MA, and London, UK: Harvard University Press.

Merchant, Carolyn. 1980. *The Death of Nature: Women, Ecology, and the Scientific Revolution*. San Francisco, CA: Harper and Row.

Murdock, Maureen. 1990. *The Heroine's Journey*. Boston, MA: Shambhala.
Panofka, Theodor. 1842. *Terracotten des Königlichen Museums zu Berlin*. Berlin: Reimer.
Peyrard, François. 1793. *De la Nature et des ses Lois*. Paris: Chez Louis.
Pliny the Elder. 77–78 AD. *Naturalis historia*. English Trans. Bostock, John, and Henry T. Riley. 1855. *The Natural History of Pliny*. London: Taylor and Francis.
Quandt, Guilelmus. 1995. *Orphei Hymni*. Berlin: Weidman.
Raine, Kathleen, and George Mills Harper, eds. 1969. *Thomas Taylor, the Platonist: Selected Writings*. Princeton, NJ: Princeton University Press.
Ripa, Cesare. 1593. *Iconologia overo descrittione dell'imagini universali cavate dall'antichità et da altri lvoghi*. Rome: Gli heredi di Giov. Gigliotti.
Roller, Lynn Emrich. 1999. *In Search of God the Mother: The Cult of Anatolian Cybele*. Berkeley and Los Angeles, CA: University of California Press.
Røstvik, Camilla Mørk. 2020. Mother nature as brand strategy: Gender and creativity in tampax advertising 2007–2009. *Enterprise & Society* 21 (2): 413–452.
Sambucus, Joannes. 1564. *Emblemata cum aliquot nummis antiqui operis*. Antwerp: Christophe Plantin.
Sheldrake, Rupert. 1991. *The Rebirth of Nature: The Greening of Science and God*. New York, NY: Bantam Books.
Smolak, Kurt. 1987. Der Hymnus des Mesomedes an die Natur. *Wiener Humanistische Blatter* XXIX: 1–14.
Stano, Simona. 2019. The corporeal meaning of language: A semiotic approach to musical glossolalia. *Semiotica* 229: 69–85.
Taylor, Timothy D. 2007. *Beyond Exoticism: Western Music and the World*. Durham-London: Duke University Press.
Toland Frith, Katherine. 1995. Advertising and mother nature. In *Feminism, Multiculturalism, and the Media: Global Diversities*, ed. Angharad N. Valdivia, 185–196. London-New Delhi: SAGE Publications Inc.
Trzaskoma, Stephen M., R. Scott Smith, and Stephen Brunet, eds. (with an Appendix on Linear B Sources by Thomas G. Palaima) 2016. *Anthology of Classical Myth: Primary Sources in Translation*. Indianapolis, IN: Hackett Publishing Company.

## Filmography

AA.VV. 2004–2012. *Desperate Housewives*, 8 seasons, 180 episodes. United States.
ACCO. 1970s. *Chiffon margarine* commercial, 29″. United States. https://www.youtube.com/watch?v=ijVijP-CDVI&t=29s. Accessed 30 Dec 2022.
Barton Thomas, Dave. 2008. *A Miser Brothers' Christmas*, 49'. United States-Canada.
Cameron, James. 2009. *Avatar*, 162'. United States-United Kingdom.
Cameron, James. 2022. *Avatar: The Way of the Water*, 192'. United States.
Cenci, Roberto. 2007–2019. *Ciao Darwin*, seasons 5–8. Italy.
Conservation International. 2014. *(Julia Roberts is) Mother Nature*, 1' 57″. https://www.conservation.org/nature-is-speaking/julia-roberts-is-mother-nature. Accessed 11 Dec 2022.
Frankel, David. 2006. *The Devil Wears Prada*, 109'. United States.
Glaceau. 2009. *Vitamin Water Mother Nature* campaign. United States.
Hathcock, Bob, George Gordon, Carl Urbano, and Rudy Zamora. 1982. *All Hallows' Eve—The Smurfs*, season 1, 22' 42″. https://smurfs.fandom.com/wiki/The_Smurfs_Springtime_Special?file=The_Smurfs_Springtime_%25E2%2580%25A2_Episode_%25E2%2580%25A2_The_Smurfs. Accessed 11 Dec 2022.
Howley, John. 1993 [1989]. *Happily Ever After* [originally released as *Snow White: The Adventure Continues* in the Philippines], 75'. United States. Trailer: https://www.youtube.com/watch?v=9Y2xNuaxgWw. Accessed 11 Dec 2022.

Lantz, Walter. 1999–2022. *The New Woody Woodpecker Show*. 3 seasons, 53 episodes. United States.
Lembeck, Michael. 2002. *The Santa Clause 2: The Mrs. Clause*, 104'. United States.
Lembeck, Michael. 2006. *The Santa Clause 3: The Escape Clause*, 92'. United States.
Leo Burnett. 2007–2009. *"Outsmart Mother Nature"* campaign—Tampax. Europe and US.
Musker, John, and Ron Clements. 2006. *Moana* (or *Vayana*, or *Oceania*), 107'. United States.
Patterson, Ray (supervising director), Oscar Dufau, George Gordon, Carl Urbano, John Walker, and Rudy Zamora. 1983. *All Hallows' Eve—The Smurfs*, season 3. 22' 36″. https://smurfs.fandom.com/wiki/All_Hallows%27_Eve?file=All_Hallowseve_%25E2%2580%25A2_Full_Episode_%25E2%2580%25A2_The_Smurfs. Accessed 30 Nov 2022.
Posten. 2022. *Father Christmas and Mother Earth*. Norway, 3' 56″. https://www.youtube.com/watch?v=cRM5PpsAH4g. Accessed 16 Dec 2022.
Rankin, Arthur Jr., and Jules Bass. 1974. *The Year Without a Santa Claus*, 51'. United States.
Recchia, Beppe. 1998–2003. *Ciao Darwin*, seasons 1–4. Italy.
Smith, Paul, Alex Lovy, and Sid Marcus. 1957–1972. *The Woody Woodpecker Show*. 5 seasons, 113 episodes. United States.
Wedge, Chris. 2013. *Epic*, 102'. United States.

# Chapter 4
# Between Natural and Cultural Catastrophes: A Look at Apocalyptic Collective Imaginaries

**Abstract** This chapter deals with the analysis of collective imaginaries of catastrophic events, ranging from fictional discourses, with a specific focus on cinema, to non-fictional representations, especially as related to the so-called "Anthropocene", environmental discourses and the Covid-19 pandemic. The analysis of fictional imaginaries points to the interconnection between "natural" disasters or calamities, on the one hand, and "cultural" catastrophes, on the other hand: while the former seem to be at least partially related to ecological or social dislocations caused by (sometimes unconscious, sometimes irresponsible) human actions, the latter can be conceived as natural entropic degenerations of complex organisations. However, as non-fictional discourses clearly show, a dichotomic view of the relation between nature *versus* culture in the conception and communication of catastrophic events is far from be abandoned.

## 4.1 The Catastrophe as a "Critical Point"

The term "catastrophe" is particularly interesting from a semantic point of view. Etymologically, it recalls the idea of an "overturning", as it derives from the ancient Greek lexeme *καταστροφή* (*katastrophḗ*), composed of *katá*, "down, against", and *stréphein*, "to turn". In this sense, it has also been largely used in the field of so-called "catastrophe theory", which originated with the work of the French mathematician René Thom (see in particular Thom 1973, 1983). Without going into the specificities of this theory, it is enough to highlight—following Cassone et al. (2018: 11–13)—that it recalls the idea of a "critical point", which could be broadly described as the final point of a phase of equilibrium, opening the way to change and transformation.[1] This view also recalls the concept of "a sudden transformation of things" [*subita rerum commutation*] described by Erasmus in his *Adagia* (1500–1536; Engl. Trans. 1989: 178), where he depicted the catastrophe as "the outcome of anything" (*Ibid.*:

[1] Such an idea was further developed by Petitot (1979, 1985), in the aim to merge the theory of catastrophe with Greimassian semiotics. A special issue of *Actes Semiotiques* devoted to the intersections between these fields also derived from the collaboration between Petitot and Thom (1983).



S. Stano, *Critique of Pure Nature*, Numanities - Arts and Humanities in Progress 26,
https://doi.org/10.1007/978-3-031-45075-4_4

177), therefore insisting on its relation with the idea of the "end". At the same time, however, it points to the relation between such an end and a new beginning deriving from it, thus highlighting its "engendering" potential.

What is more, if we look at contemporary dictionaries, another element is added. The Oxford Dictionary (2022), for instance, describes the catastrophe as "an event causing great and usually sudden damage or suffering; a disaster". Accordingly, the Cambridge Dictionary (2022) defines it as "a sudden event that causes very great trouble or destruction", and the Merriam-Webster (2022) depicts it as "a momentous tragic event ranging from extreme misfortune to utter overthrow or ruin" (which also partially evokes the idea of an overturning). A marked dysphorisation has thus evidently emerged over time, as suggested by the association of the term with expressions such as "damage", "suffering", "disaster", "great trouble", "destruction", "tragic event", "extreme misfortune", "utter overthrow", and "ruin", as well as by its common use (e.g. "That's a catastrophe!").

Moreover, a peculiar relation with the "semantic halo" associated with nature (see Chap. 2) has developed, as demonstrated by the fact that all the considered definitions refer to the "environment" and/or to "ecology" when providing examples of typical uses of the word: "an *environmental* catastrophe" (Oxford Dictionary 2022, my emphasis), "They were warned of the *ecological* catastrophe to come" (Cambridge Dictionary 2022, my emphasis), "Deforestation and erosion can lead to an *ecological* catastrophe" (Merriam-Webster 2022, my emphasis). In fact, one of the most pressing concerns of contemporary societies is related to a series of "critical points" affecting the environmental dimension, especially as related to climate change, population growth and pollution.

Four main aspects therefore emerge as fundamental in defining catastrophic events nowadays: the idea of an "overturning"; a *terminative* characterisation,[2] which is nonetheless immediately related to a "new beginning" (i.e. an *inchoative* characterisation[3]); a dysphoric connotation, which sees them as disasters, ruins, namely destructive and tragic events; and a particular link with the natural domain, intended as the environmental and ecological domain. In order to further explore these aspects, and in particular the last of them, the following paragraphs will deal with the analysis of relevant examples of discourses on catastrophic events, as related to both fictional and non-fictional texts.

---

[2] The reference is here to the theory and terminology of aspectualisation, as described by Greimas and Courtés: "Terminativeness is an aspectual seme indicating the completion of a process …; when it is recognized, one can presuppose the existence of the entire configuration. At the level of surface semiotic syntax, *terminativeness* may indicate the realization of a doing" (Greimas and Courtés 1979; Engl. Trans. 1982: s.v. "terminativeness").

[3] "Inchoateness is an aspectual seme which shows the beginning of a process … its appearance in the discourse allows the realization of the entire series to be foreseen or expected" (*Ibid.*: s.v. "inchoateness").

## 4.2 Catastrophic Scenarios and the Nature/Culture Dilemma

Based on the aspects described above, the catastrophe has been generally associated with the idea of the "end of the world"—with a markedly anthropocentric conception of the term "world", which usually stands for human civilisation and/or planet Earth (Terrone 2018: 89).

In fact, such a topic has traditionally played a crucial role in the collective imaginary of catastrophes. From mythological and religious eschatology to modern ecological policies, from science fiction to contemporary journalism, representations of catastrophic events determining the destruction of our planet and the extinction of humankind are abundant.

Various sacred traditions—as varied as African shamanism and New Age cults, also embracing the religious matrices of Western Asia that gave rise to Judaism, Christianity and Islam (Ruthven 2015)—, for instance, envision the Apocalypse (intended, according to the etymology of this term,[4] as a prophetic "revelation", a "disclosure" of previously inaccessible truths and knowledge) in association with world cataclysms, disasters and devastating wars—that is to say, with an "overturning"(i.e. a catastrophe) of both the natural and the cultural order. As I observed in Stano (2018), similar representations characterise also Native American beliefs, which for instance refer to the appearance of a blue *katchina* star and a red star foretelling a universal war among men (see Waters 1993), as well as the Norse myth of *Ragnarök*, the "twilight of the gods", which is marked by a war between Scandinavian divinities, in turn anticipated by natural catastrophic events such as a 3-year long extremely cold winter and the submersion of world in water (see Bellows 2004).

Another interesting example in this sense can be found in the Bible, and more specifically in the popular flood narrative reported in the *Book of Genesis*:

> So God said to Noah, "I am going to put an end to all people, for the earth is filled with violence because of them. I am surely going to destroy both them and the earth. So make yourself an ark of cypress wood; make rooms in it and coat it with pitch inside and out. This is how you are to build it: The ark is to be three hundred cubits long, fifty cubits wide and thirty cubits high. Make a roof for it, leaving below the roof an opening one cubit high all around. Put a door in the side of the ark and make lower, middle and upper decks. I am going to bring floodwaters on the earth to destroy all life under the heavens, every creature that has the breath of life in it. Everything on earth will perish. But I will establish my covenant with you, and you will enter the ark—you and your sons and your wife and your sons' wives with you. You are to bring into the ark two of all living creatures, male and female, to keep them alive with you (*Genesis* 6, 13–19). … And it came to pass after seven days, that the waters of the flood were upon the earth. In the six hundredth year of Noah's life, in the second month, the seventeenth day of the month, the same day were all the fountains of the great deep broken up, and the windows of heaven were opened (*Genesis* 7, 10–12).

In this case too, a subversion of the cultural order (resulting from the state of generalised violence caused by mankind) is related to an overturning of the natural

[4] Derived from the Greek ἀποκάλυψις (*apokalupsis*), which is composed of *apó*, "off, away from", and *kalýptein*, "to cover, conceal".

order (the universal flood), which in turn results in the destruction of such a subverted culture, and the origin of a new order, based on the alliance between God and Noah (i.e. the uniquely righteous man in an era of exceptional wickedness in the world, who therefore represents an "alternative culture").

It must be noted that the combination of natural and cultural elements leading to the end of the world is not a prerogative of ancient myths and sacred texts. In fact, it also plays a crucial role in scientific theories, ranging from lethal pandemics capable of annihilating the human population to the effects of pollution on the biosphere and the ecosystem, from glaciation to extreme astronomical events, such as the collision of large meteoroids or other celestial objects against the Earth (which would kill most life forms within several kilometres, also causing the collapse of temperatures and the interruption of the food chain, in addition to violent earthquakes and tsunamis), as well as nuclear wars, etc. These theories, along with mythical and religious eschatology, have profoundly affected our collective imaginaries, not only as related to literary and audiovisual fiction, where their representations abound, but also non-fictional discourses and debates concerning everyday life (especially when a catastrophic event such as the Covid-19 pandemic came to actually threaten it, or other "critical points" concerning the environment emerged more clearly). I will focus on some relevant cases studies in the following paragraphs, dealing specifically with cinema, on the one hand, and environmental campaigns, digital communication and journalism, on the other hand.

## 4.3 Fictional Apocalyptic Collective Imaginaries: The Case of Cinema

Catastrophic cinema includes several interesting examples, from the 1933 American apocalyptic science fiction movie *Deluge*, directed by Felix E. Feist and based on the homonymous novel by Sidney Fowler Wright published in 1928, which depicts a series of unexplained natural disasters (e.g. a sudden eclipse of the sun, tremendous earthquakes and major tsunamis) erupting around the world and destroying human civilisation, to the recent American apocalyptic black comedy *Don't Look Up* (2021), written and directed by Adam McKay from a story he co-wrote with David Sirota, which is based on the discovery of a massive comet approaching the Earth and threatening human life. In between, a number of other popular movies can be mentioned: consider, for instance, *The Day After Tomorrow* (Emmerich 2004), where human pollution inexorably leads to catastrophic environmental phenomena leaving but a few survivors; *2012* (Emmerich 2009), which recalls the episode of the universal flood described in the Bible, albeit with significant variations; and also *WALL·E*, a (2008) computer-animated science fiction film directed by Andrew Stanton, where humanity has already abandoned the Earth, which seems deprived of any form of biological life[5] (until WALL·E finds a plant) because of pollution, to move to space on giant "star ships".

Despite their peculiarities, these examples share interesting thematic aspects, which makes them emerge as variants of two main "matrices" (in the sense described by Lévi-Strauss 1958), or general models, which problematise the complex relationship between "nature" and "culture". Such matrices, and their implications, will be described in detail in the following paragraphs,[6] based on the analysis of the cinematographic examples mentioned above.

### *4.3.1 Matrix 1: The End of the World as a "Natural" Catastrophe*

On the one hand, there are texts such as *Deluge* or *2012*, which focus on extreme *natural* factors (referred to as "Extreme Nature" in Fig. 4.1), seemingly inexplicable and completely uncontrollable, to explain the origin of the end of the world. Whether it is an eclipse, an earthquake or a universal flood, such factors do not seem to leave a chance for humanity, understood not merely in terms of human life (which somehow seems to find a way to resist, though with huge losses), but rather as related to civilisation, that is to say, in reference to the *cultural* system developed up to that moment ("Destruction of Culture 1"). However, an alternative system ("Alternative Culture 2"), recognised as more just, supportive and respectful of Nature, generally arises from such a catastrophe, replacing the annihilated Culture 1.

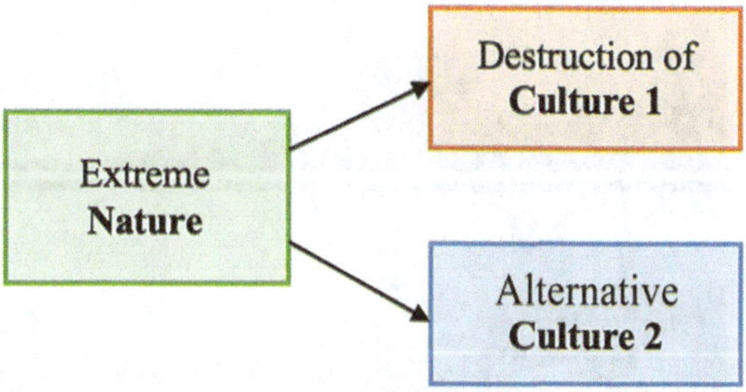


**Fig. 4.1** Matrix 1: the end of the world as a "natural" catastrophe

[5] In fact, a cockroach also appears in the movie as WALL·E's friend, distinguishing itself for a particular appetite for Kremies and an extreme physical resilience (to the extent that it actually "resuscitates" after the robot accidently crushes it). However, its inclusion is used precisely as a reference to the myth that roaches (and Twinkies, which inspired the Kremies featured in the movie) are the only things that can survive the end of the world.

[6] Which build on some aspects identified in Stano (2018), further developing them and comparing them with other relevant case studies.

The first sequences of *Deluge*, for instance, juxtapose images of uncontrollable and mysterious natural phenomena with shots of incredulous experts wondering about them and close-ups of alarming newspaper front pages referring to the inexplicability of such phenomena and the dismay and disorientation of the scientific community (see Fig. 4.2). This opens the way to the progressive decline of civilisation: not only all urban and cultural artefacts (including symbolic elements such as the Empire State Building and the Statue of Liberty in New York City) are disintegrated by seismic tremors and floods, with a sort of return to a primordial "natural" state forcing people to live in caves and huts arranged where possible; but violence, ineptitude and tyranny also spread everywhere, with a total loss of the sense of common good.

It is only towards the epilogue of the movie, and thanks to the exceptional action of exemplary men such as Morris Webster (personified by actor Sidney Blackmer), who distinguished himself from the others, that a possibility of renaissance arises, in view of "a finer, cleaner civilisation" (as the protagonist himself says, 63′ 49″), which learns from the past and what has been destroyed to create an alternative, and better, culture.

**Fig. 4.2** Screenshots from the movie *Deluge* (Feist 1933)

**Fig. 4.3** Screenshots from the trailer of the movie *2012* (Emmerich 2009)

A similar pattern can also be found in the movie *2012*, which puts a particular emphasis on the destruction (always due to extreme natural events, see Fig. 4.3, frames 1–3) of cultural emblems around the world (e.g. the White House and the Washington Monument in the United States, the Sistine Chapel in Italy, the statue of Christ the Redeemer in Rio de Janeiro, etc., see Fig. 4.3, frames 4–7) and the threat of the disappearance of any form of humanity, symbolised by a sudden interruption of the presidential speech at the very moment when he invokes religious faith and hope (see Fig. 4.3, frame 8).

In this case too, however, an alternative cultural model (which finds expression in the White House scientific consultant Adrian Helmsley's final speech, aimed at convincing the privileged passengers of the shuttles designed to survive the universal flood to all the people remained outside of them, thus saving them from a certain death) seems to be able to survive such natural disasters, opening the way to new horizons, as the epilogue of the movie remarks, when the shuttles approach the coasts of the African continent for a "new life" (see Fig. 4.3, frame 9).

### 4.3.2 *Matrix 2: The End of the World as a "Cultural" Catastrophe*

Other movies seem to rely on a seemingly different model, where catastrophes are no longer triggered by nature, intended as an uncontrollable force independent from human activity, but rather by degenerated cultural systems ("Dysphoric Culture 1"),

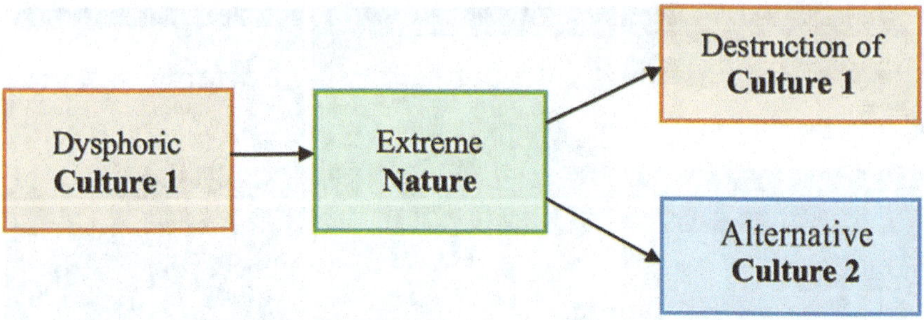


**Fig. 4.4** Matrix 2: the end of the world as a cultural catastrophe

whose selfishness and disrespect of the former provokes extreme responses from the former (according to the relations described in Fig. 4.4).

This is the case, for instance, of *The Day After Tomorrow*, as exemplified by the very first scenes of the movies: when the world-renowned climatologist Professor Jack Hall (actor Dennis Quaid) predicts an imminent planetary glaciation at a world conference on global warming, he is strongly opposed by the vice-President of the United States (representative of the "Dysphoric Culture 1"), who favours economic logics over the scientist's warning. However, Hall's predictions do not take long to come true, with extreme and uncontrollable natural events beating planet Earth (see Fig. 4.5) and causing the destruction of "Culture 1", thus opening the way to an alternative, more responsible "Culture 2".

**Fig. 4.5** Screenshots from the trailer of the movie *The Day After Tomorrow* (Emmerich 2004)

Similarly, in *WALL·E*, the death of nature on Earth is presented as a result of a dysphoric culture, whose widespread consumerism, corporate greed and environmental neglect turned the planet into a garbage-strewn wasteland, forcing humanity to evacuate to space. An alternative cultural model interestingly arises with WALL·E, the last remaining active trash-compacting robot (or Waste Allocation Load-Lifter:

Earth-Class, as the acronym suggests) left on Earth by humans to clean it up, which in time developed a proper personality (i.e. an alternative "culture"), salvaging and collecting objects from garbage. This is precisely what leads the robot to save from destruction a living seedling, thus activating the adventurous path that will conduct humanity to (re-)discover a cultural system based on pre-consumeristic values and their very "nature"—as it is also interestingly remarked by the tension between the humans' struggle to recover the "natural" movements of their body (inhibited by years of physical inactivity on the spaceship) and the resistance opposed by the technological prostheses that have expanded, but also increasingly controlled and impoverished, their capabilities over centuries.

### *4.3.3 A Unifying Model: Nature as Culture*

The models presented above only apparently—and partially—differ from each other. On closer inspection, in fact, the two catastrophic dynamics described in the previous paragraphs refer to genres that have always been interconnected in a sort of "heterogenesis of ends" (Gattei 2013): as both Diamond (2005) and Jacobelli (2013) argue, in fact, if natural disasters seem to be to some extent attributable to ecological or social imbalances caused by (sometimes unconscious, sometimes irresponsible) human actions, cultural catastrophes in turn seem in some ways conceivable as "natural" entropic degenerations of complex organisations. It is in this sense that the collective representations of the end of the world call into question the crucial issue of "multinaturalism"—or, better, "internaturality" (as described in Chap. 2)—, questioning the heuristic and ontological value of the categories of "nature" and "culture" themselves. Indeed, the matrices described above can be reasonably traced back to a more general common model, which can be represented through the scheme shown in Fig. 4.6.

Such a model encompasses both a movement of "naturalisation" of nature, which sees this as completely independent from human action, and a process of "culturalisation" of nature, which generally relies on the figure of an excessive and reckless

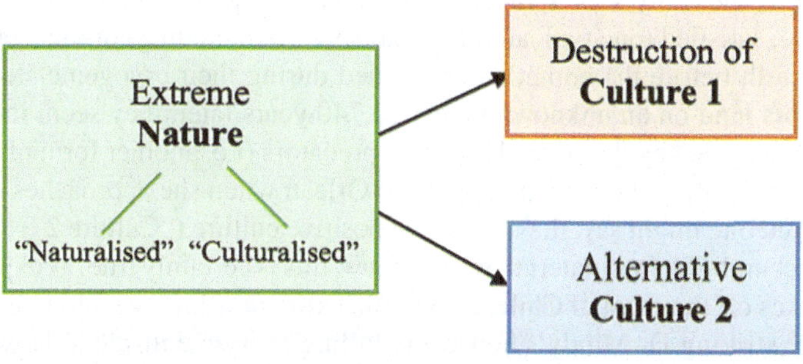


**Fig. 4.6** Matrix 3: Integration of matrices 1 and 2

exploitation of natural resources by a dysphoric cultural system (coinciding precisely with the destroyed "Culture 1"). In any case, nature is intrinsically linked to culture, as it is experienced and conceived through it.

This aspect is interestingly recalled in a more recent movie representing the end of the world: *Don't Look Up*. Having discovered that a giant comet is going to collide with the Earth in about six months, causing a planet-wide extinction event, PhD student Kate Dibiasky (actress Jennifer Lawrence) and professor Dr Randall Mindy (actor Leonardo Di Caprio) are taken to the White House by NASA Planetary Defense Coordination Office head Dr Clayton "Teddy" Oglethorpe (Rob Morgan) to inform the President of the United States, Janie Orlean (Meryl Streep), and her son and Chief of Staff Jason (Jonah Hill) about this. After facing the politicians' indifference, they decide to leak the news to the media, participating in a morning talk show. However, the topic is treated frivolously by hosts Jack Bremmer (Tyler Perry) and Brie Evantee (Kate Blanchett), and Dibiasky is even publicly mocked for insisting on the gravity of the situation. It is only later, because of the need to distract people from the bad publicity concerning her party, that the President confirms the threat and announces a project to strike and divert the comet using nuclear weapons. However, after the mission successfully launches, it is forcefully aborted for economic reasons, despite the scientists' warnings that the alternative solution proposed by Peter Isherwell (interpreted by actor Mark Rylance), the CEO of BASH (a major telecommunications company), has not received any scientific validation. Even when the comet becomes visible from Earth, and the researchers start a protest campaign inviting people to "Just Look Up", calling on other countries to conduct comet interception operations, the President responds with the anti-campaign "Don't Look Up". As the catastrophe is approaching, due to the (predictable) failure of BASH's mission, Orlean and others in her elite circle board a sleeper spaceship designed to find another planet where to live, while Dr Mindy, despite being offered two places on the ship, decides to remain on Earth and spend his last evening with his friends and family.

At first glance, it might seem that the movie overthrows the model described above, making the dysphoric culture ("Culture 1") that allegedly ignores the imminent catastrophe, favouring political and economic interests over scientific warnings, survive, and fating the alternative—positive—culture of science ("Culture 2") to death. However, as a mid-credits scene reveals, the very epilogue of the story contrasts this: not only 42% ("much better than anticipated", as Isherwell remarks in the naïve, satisfied tone typical of his character upon landing) of the 2,000 people who left Earth before the comet's impact died during their cryogenic sleep, but as the survivors land on an unknown planet 22,740 years later, they seem inescapably destined to be killed by the several bird-like predators (i.e. another form of "Extreme Nature") surrounding them, as it happens to Orlean when she approaches one of the animals. Yet, one might say, the alternative positive culture ("Culture 2") also seems fated to succumb. From a material point of view, this is certainly true: as expected, the comet strikes off the coast of Chile, causing an extinction-level event, with a gigantic shockwave striking Dr Mindy's house and killing everyone inside it. However, just before dying, these characters find their full redemption, becoming able to actually "look up" at the broader system they are part of, where nature and culture are no longer separated, but inextricably connected to each other. In fact, while "Culture 1" cannot but assist to its own failure and escape, Dr Mindy is able to reunite with his family and friends (and also with his real self). They distance themselves from the corruption and nonsense of "Culture 1" (as the soundtrack interestingly remarks), as they try to "face what … is coming with … open hearts of acceptance" (as one of the characters says in a prayer), fully reconnecting with a nature they no longer see as an alterity to be measured or contrasted, but rather as the realm they also form part of.[7] Moreover, after the collision, the boundaries between nature and culture seem to completely disappear: both the elements generally ascribed to the former (e.g. pieces of the terrestrial crust, eradicated trees, shattered whales, etc.) and those attributed to the latter (e.g. cars, Dibiasky's smartphone, ruins of buildings, the sign of a restaurant,[8] the Bull of Wall Street,[9] etc.) float around what was planet Earth, while the ship hosting the representatives of "Culture 1" escapes towards "new horizons".

## 4.4 From Fictional to Non-fictional Apocalyptic Collective Imaginaries: A Never-Ending Dichotomy?

The dynamics highlighted above are not exclusive to fictional discourses, but largely characterise non-fictional representations too.

This is evident, for instance, in the contemporary debate concerning the so-called "Anthropocene", an idea arisen between the 1980s and 1990s[10] and then

[7] In fact, the movie interestingly juxtaposes images of these characters eating, talking, praying, and eventually dying, with images of moving whales, bees, bears, deer, fish, trees, etc., accompanied by a melodic and calm soundtrack (which contrasts the noise produced by BASH's mission and the chaos generated among people when destruction is finally seen as unavoidable), until everything goes to silence and darkness (i.e. death). On the other hand, even after the destruction of Earth caused by the meteorite, the survived representatives of Culture 1, despite entering "naked" (which is generally conceived as a symbolic reference to "Nature", although erroneously, cf. Chap. 6) the new planet where they have just landed, still maintain an evident "detachment" from nature, with a marked opposition between the latter and the culture they represent.

[8] More specifically, Bojo Mambo's, i.e. the restaurant where Kate previously had a public meltdown, announcing to the gathered patrons the President's decision not to pursue a mission that would break up the comet to further enrich some billionaires.

[9] A clear reference to economic interests.

[10] In fact, biologist Eugene F. Stoermer (see Grinevald 2007: 243) claimed to have started using the term in the 1980s, but never formalized it until Paul Crutzen contacted him. Steffen et al. (2011) remarked that other authors (such as Nisbet 1991) also explored similar concepts about the same time, although not using that specific term, and reported an interesting quote from a volume on *Global Warming* published in 1992 by Andrew C. Revkin: "Perhaps earth scientists of the future will this new post-Holocene period for *its causative element*—for *us*. We are entering an age that might someday be referred to as, say, the *Anthrocene* [sic]. After all, it is a geological age of *our own making*" (in Steffen et al. 2011: 843, my emphasis). Moreover, the increasing influence of mankind on the environment is central to the idea of *noosphere*, lit. "the world of thought", as described by scholars such as Vernadsky (1991), de Chardin (1957 [1923]) and Le Roy (1927, 1928) back

popularised[11] by Paul Crutzen and Eugene Stoermer in 2000 as a reference to "the present geological time interval, in which many conditions and processes on Earth are profoundly altered by human impact" (International Commission on Stratigraphy—Anthropocene Working Group 2019[12]; cf. Crutzen and Stoermer 2000: 17), resulting in a series of ecological, or environmental catastrophes (in the sense described above). If, on the one hand, this concept has been celebrated by some (see, for instance, Pellegrino 2022) precisely as the collapse of all dichotomic views opposing Nature to Culture, the natural domain to the human sphere, and natural sciences to human sciences, it has also resulted in a further accentuation of such oppositions. In fact, as illustrated by Bonneuil (2015) and Bonneuil and Fressoz (2015; cf. Di Paola and Pellegrino 2018), four major narratives of the Anthropocene can be identified. Whereas the *post-nature* narrative proclaims the "impossibility to continuing to separate 'nature' and 'society'" (Bonneuil 2015: 24), the others—namely, the *naturalist* narrative, which points to an undifferentiated humanity uniformly concerned by and responsible for global environmental change; the *eco-catastrophist* narrative, which insists on the destruction of nature caused by the excessive consumption of resources by men, resulting in an irreversible and dangerous change in the conditions of our planet; and the *eco-Marxist* narrative, which depicts the Anthropocene as the culmination of the contradictions of capitalism—remain essentially anchored to a dichotomic view of nature as opposed to culture/society. Accordingly, a number of contemporary environmental discourses[13] maintain such an opposition, representing now a "depleted" nature (e.g. primary resources, energy and water scarcity[14]),

---

in the first decades of the 20th century (for a general review on the genesis of the term and its understanding see Shushakov 2020).

[11] Not without criticisms (see, for instance, Santana 2019).

[12] http://quaternary.stratigraphy.org/working-groups/anthropocene/ (Accessed: 22 December 2022).

[13] For some interesting analyses, see for instance (Painter et al. 2016; Allgaier 2019; Lozano and Nakayama 2022).

[14] As discussed in a number of articles and essays dealing with population growth, and especially with the need to "develop socio-economic systems and institutions that are not dependent on continued material consumption growth" (Royal Society of London 2012, in Magdoff 2013: 27), which recall particularly the above-mentioned eco-Marxist narrative. Various environmental campaigns have also highlighted this aspect, as exemplified by the World Wildlife Fund (WWF)'s poster (Uncle Grey 2007) focusing on the continuing destruction of the rain forest, which featured a Tarzan-like man throwing himself off a rope from a tree. While half of the picture (and more specifically, the part where the man comes from) hosts a green, thriving forest, the other half (where the man is about to land) is desert and lifeless. Moreover, the sky is covered by partly black clouds, with a resulting dysphoric effect, which is thus projected on the main message of the advertisement, reported at the bottom of the poster: "15 $km^2$ of rain forest disappears every minute".

now an "altered" nature (e.g. climate change and global warming[15]), now a "disfigured" nature (e.g. as a result of the pollution and damage caused by plastic and other substances[16])—in any case a reality ontologically different and separate from the human and socio-cultural world, which has reached a "critical point" precisely because of the activities of the latter.[17]

What is more, this ideology has become dominant also as related to catastrophic events not directly related to such environmental concerns, as exemplified by the emblematic case of the Covid-19 pandemic. Since its first outbreak in Wuhan (China) in late 2019, in fact, the Coronavirus disease—as well as the major catastrophe it caused—has been the object of a number of discourses entailing the link between nature and culture. In fact, while most scientists agree that the virus is the product of natural evolution (see, for instance, Andersen et al. 2020), other sources (mainly coming outside of recognised scientific institutions, but nonetheless largely spread in global mediascapes[18]) still maintain that it was a laboratory construct, or in any case a purposefully manipulated virus, thus highlighting its cultural origin. In any case, a major relation with contemporary culture—interestingly often referred to precisely as the "Anthropocene"—was stressed since the beginning, as exemplified by its descriptions in terms of an effect of "our troubled relationship with the 'natural'

---

[15] Consider, for instance, the campaign "Roomies from the Wild" by Artificial Group (2023) for WWF, whose main claim that "The environmental impact of humankind on nature is catastrophic" finds expression in visual paradoxes (e.g. an otter swimming in a bathtub, a wild boar with her cubs lying on a bed, a lynx on a sofa facing the worried gaze of a more domestic cat, etc.) and verbal warnings concerning the "catastrophic" effects that might result from such an alteration of nature by man ("If wetlands disappear, can she move in with you?", "If forests disappear, can they/he move in with you?"), and therefore an invitation to "Protect [their] natural habitat instead". The famous WWF's poster featuring a melting scoop of ice cream visually recalling the Earth on top of a cone (VVL BBDO Belgium 2005) is also particularly interesting in this sense, as it shows a deformed, and exhausted, planet (as a result of global warming), mixing the figure of the "altered" Nature with that of the "depleted" Nature considered above—and evoking especially the above-mentioned eco-catastrophist and naturalist narratives, exactly like the other considered examples.

[16] An emblematic example in this sense is the thought-provoking cover published in 2018 by National Geographic magazine as part of its multiyear campaign "Plastic or Planet?", intended to raise awareness about the global plastic trash crisis, with a particular emphasis put on the naturalist narrative described above: a picture by photographer Jorge Gamboa showed a plastic bag floating in the sea, with its upper part sticking out of the water, recalling the figure of an iceberg and thus becoming a metaphor for the huge "disfiguration" plastic is causing to our planet. Similarly, a poster released by the Mediterranean Association to Save the Sea Turtles (MEDASSET) juxtaposed images of jellyfish and upside-down plastic bag, stating "You see the difference. A turtle does not" (with an implicit reference to the catastrophic effect caused by such a disfiguration of nature—i.e. the ocean and what it contains—on nature itself—i.e. turtles). Finally, a mention should be made to the campaign against plastic by UNICEF (in collaboration with Helwan University 2019), which played on eidetic and chromatic configurations to "enclose" a sea turtle into a water tank (with a disfiguration of the exteriority of her body), or—in other posters—to denounce the ingestion of plastic by seahorses and crabs (with a disfiguration of the interiority of their body).

[17] In this sense, it is particularly interesting to consider the increasing number of environmental demonstrative actions involving artworks, monuments and other cultural artefacts, which are generally endowed with particular symbolic meanings (see Bessette and Bessette 2022).

[18] For a detailed summary and discussion, see in particular Lewandowsky et al. (2022).

world" (Castro 2020) and "a shocking reminder that our control of nature, on which modern life rests, is more fragile than we like to think" (Tooze 2020).

Moreover, when the Covid-19 became a global pandemic, requiring mitigation measures such as travel restrictions, quarantines and lockdowns, a real "catastrophe" (in the etymological sense of an "overturning") arose, further enhancing the opposition between nature and culture. As Camilla Pietrabissa highlighted, in the months following the beginning of the pandemic, visual and textual references to urban void became dominant:

> When, since the early weeks of March, European cities began to close down too, newspapers started to publish photographs of empty squares and streets. This category includes a variety of composition schemes, from pictures taken by drones, to pictures taken from street level which, through side-angle perspectives, emphasise the (lack of) human passage on the ground rather than the vision of architecture. Manholes, benches, pedestrian dividers, the pattern of pavements, and other kinds of urban decorations and props took centre stage. As the ground occupied a larger portion of the composition, the visibility of elements of urban design manifested the absence of people (Pietrabissa 2020: 17).

While initially this caused a sense of emptiness even on an emotional level (Cati 2020), soon "the search for an otherness, capable of redefining the boundaries of existence from a new relationship between Man and Nature" (*Ibid.*: 77) prevailed. In fact, images and videos revealing the re-appropriation of urban spaces by ducks, rabbits, swans, flamingos, deer, bears and whales, and even simpler natural organisms (such as the grass flourishing among cobblestones in Rome[19]) spread like wildfire throughout the social media, documenting what was acclaimed as a miraculous "revenge" of nature, its unexpected "return" (see Peverini 2020). As a consequence, a number of articles and online posts appeared in print and online newspapers, with titles openly celebrating such an idea: "Nature reclaims public spaces during lockdown"[20] (Reuters 2020), "Nature and the Coronavirus: As Human Continue Lockdown, Wildlife Creeps Back In"[21] (wbur 2020), "Con le attività umane ferme o fortemente rallentate la natura riconquista i suoi spazi"[22] [*With human activities stopped or strongly slowed down, nature regains its spaces*[23]] (Corriere della Sera 2020), "Lo stop per il coronavirus ha riportato la natura in città. Adesso sta a noi trattenerla"[24] [*Coronavirus closures have brought nature back to cities. Now it is up to us*

[19] See, for instance, the video reposted by *Il Messaggero* in the article "Piazza Navona, l'erba cresce 'indisturbata' tra i sampietrini" [*Piazza Navona, the grass grows 'undisturbed' among the cobblestones*, my translation], https://www.ilmessaggero.it/video/roma/roma_piazza_navona_erba_cresce_sampietrini-5148163.html (Accessed: 19 November 2022).

[20] https://www.reuters.com/news/picture/nature-reclaims-public-spaces-during-loc-idUSRTX7EJ1H (Accessed: 19 November 2022).

[21] https://www.wbur.org/onpoint/2020/04/29/wildlife-coronavirus-animals-climate (Accessed: 19 November 2022).

[22] https://www.corriere.it/animali/bonnie-e-co/notizie/coronavirus-ci-insegna-perche-siamo-diversi-altri-animali-f307f2a2-70db-11ea-a7a2-3889c819a91b.shtml (Accessed: 19 November 2022).

[23] All translations of the titles are mine.

[24] https://www.open.online/2020/03/19/lo-stop-per-il-coronavirus-ha-riportato-la-natura-in-citta-adesso-sta-a-noi-trattenerla/ (Accessed: 19 November 2022).

*to make it stay*] (Open 2020), "La naturaleza se abre paso por el mundo tras el confinamiento"[25] [*Nature makes its way through the world after confinement*] (Cadena Ser 2020), "La naturaleza salvaje reconquista la ciudad por el coronavirus"[26] [*Wild nature reconquers the city as a result of Coronavirus*] (National Geographic España 2020), "Coronavirus. Les humains sont confinés, la nature reprend ses droits"[27] [*Coronavirus. Humans are confined, nature takes back its rights*] (Ouest-France 2020), "Confinement: le retour de la nature"[28] [*Lockdown: the return of nature*] (franceinfo 2020), "Wie das Coronavirus die Natur wiederbelebt: Menschen bleiben zu Hause, Hirsche spazieren auf der Straße"[29] [*How the Coronavirus is reviving nature: people stay at home, deer roam the streets*] (Der Tagesspiegel 2020), "Coronavirus: Die Rückkehr der Natur"[30] [*Coronavirus: the return of nature*] (WEB.DE 2020).

While the forced containment of human activities and presence in the environment undoubtedly encouraged both animals and vegetables to expand beyond their usual borders, it must be noted that several appearances were proved to be fake, or at least not entirely accurate. The case of Venice is emblematic in this sense: though the decrease in boat activity evidently made the canals water clearer and cleaner, the "return of swans and dolphins" claimed by social media was not real:

> The swans in the viral posts regularly appear in the canals of Burano, a small island in the greater Venice metropolitan area, where the photos were taken. The "Venetian" dolphins were filmed at a port in Sardinia, in the Mediterranean Sea, hundreds of miles away (Daly 2020).

Whether real or not, such news and posts contributed to give origin to an emerging rhetoric modelled on an evident idealisation of nature, stereotypically conceived as an original, idyllic and somehow "sacred" state, as opposed to the dysphoric "culture" of current societies. In fact, as Teresa Castro outlined, "as satellite images revealed significant drops in air pollution across the planet, drone footage of emptied cities came to embody, for some, what the world would look like without humans—or, at least, with considerably fewer humans" (Castro 2020: 85). Extreme positions therefore arose, leading to the spread of messages such as "Corona is the cure, humans are the disease"—a creation of far-right activists, mischievously attributed to the global environmental movement Extinction Rebellion, which immediately distanced itself from it—or "humans are the virus"—a *motif* shared countless times on social

---

[25] https://cadenaser.com/ser/2020/03/19/sociedad/1584615279_488513.html (Accessed: 19 November 2022).

[26] https://www.nationalgeographic.com.es/ciencia/naturaleza-salvaje-reconquista-ciudad-por-coronavirus_15346 (Accessed: 19 November 2022).

[27] https://www.ouest-france.fr/sante/virus/coronavirus/les-humains-sont-confines-la-nature-reprend-ses-droits-6794461 (Accessed: 19 November 2022).

[28] https://www.francetvinfo.fr/sante/maladie/coronavirus/confinement-le-retour-de-la-nature_3932615.html (Accessed: 19 November 2022).

[29] https://www.tagesspiegel.de/wissen/menschen-bleiben-zu-hause-hirsche-spazieren-auf-der-strasse-5737576.html (Accessed: 19 November 2022).

[30] https://web.de/magazine/wissen/natur-umwelt/coronavirus-rueckkehr-natur-34552246 (Accessed: 19 November 2022).

media during lockdown, especially in connection with the images described above, and also appeared in scholarly literature.[31] And even beyond these extremisms, it is impossible not to notice an increasingly marked tendency to praise such a "revenge" of nature—as demonstrated by the emblematic description provided in the section devoted to Nature (significantly entitled "Nature's Rebound"[32]) on the permanent archive of the Coronavirus pandemic hosted on the multimedia platform *The COVID-19 Visual Project*:

> In a planet that has been deserted by humans for weeks, *nature has resumed its course*: blue sharks, whales, roe deer have come out of hiding. Nice to meet you, we are the humans. We had only seen each other from afar, we hold strange objects in our hands and you are on camera. Not the wild boars though, they had already tried to conquer the city. Will we be able to respect each other more now? (my emphasis).

The pandemic, in other terms, has been largely seen as a "catastrophe" marking the end of a dysphoric culture ("Culture 1") thought to have imposed itself over Nature (which has thus responded to it, resulting in the pandemic, and has then flourished again as such a culture was forced to stop). Hence, exactly as most environmental discourses and narratives of the Anthropocene, it seems to recall the dynamics identified in fictional texts. Moreover, it takes them to the extremes, prefiguring a sort of cultural annihilation—which cannot but find expression in a markedly dystopic scenario. In fact, unlike the cases analysed above, the pandemic discourses do not seem to prefigure any alternative euphoric culture (i.e. "Culture 2"), rather opting for a denial of culture as such, that is to say, a sort of annihilation of man himself (or, at least, of his activities) in view of a "re-naturalisation" of our planet.[33]

Once again, in conclusion, nature is seen as an alterity, as something that is not only separate from, but also opposite to culture, perpetrating a dualistic vision that fails to recognise man himself as part of the natural world—as he, in fact, is.

## References

Allgaier, Joachim. 2019. Science and environmental communication on youtube: Strategically distorted communications in online videos on climate change and climate engineering. *Frontiers in Communication* 4: art. 36.

Andersen, Kristian G., Andrew Rambaut, W. Ian Lipkin, Edward C. Holmes, and Robert F. Garry, 2020. The proximal origin of SARS-CoV-2. *Nature Medicine* 26: 450–452. https://doi.org/10.1038/s41591-020-0820-9. Accessed 3 Dec 2022.

[31] The paper "Coronavirus lockdown helped the environment to bounce back" (Arora et al. 2020), for instance, reports as its main outcome that "although coronavirus vaccine is not available, coronavirus itself is earth's vaccine and us humans are the virus".

[32] https://covid19visualproject.org/en/chapter/la-rivincita-della-natura/6 (Accessed: 19 November 2022).

[33] To some extent, this seems to recall the special sense of appreciation for "wilderness"—intended "as a process of place of nature's self-reassertion" (Kirchhoff and Vincenzotti 2014: 455)—that arose in the mid-1990s, taking it to the extreme.

Arora, Shefali, Kanchan Deoli Bhaukhandi, and Pankai Kumar Mishra. 2020. Coronavirus lockdown helped the environment to bounce back. *Science of the Total Environment* 742: 140573.

Bellows, Henry A., ed. 2004. *The Poetic Edda: The Mythological Poems*. Mineola, NY: Dover Publications.

Bessette, Anne, and Juliette Bessette. 2022. On environmental activism in museums. *e-flux—Notes*, December 06. https://www.e-flux.com/notes/507828/on-environmental-activism-in-museums. Accessed 10 Aug 2023.

Bonneuil, Christophe. 2015. The geological turn. Narratives of the Anthropocene. In *The Anthropocene and the Global Environmental Crisis. Rethinking modernity in a new epoch*, ed. Clive Hamilton, François Gemenne and Christophe Bonneuil, 12–31. London: Routledge.

Bonneuil, Christophe, and Jean-Baptiste Fressoz. 2015. *The Shock of the Anthropocene: the Earth, History and Us*. London: Verso.

Castro, Teresa. 2020. Of drones and the Environmental Crisis in the Year of 2020. In *Pandemic Media. Preliminary Notes Toward an Inventory*, ed. Philipp Dominik Keidl, Laliv Melamed, Vinzenz Hediger and Antonio Somaini, 81–88. Lüneburg: Meson Press.

Cati, Alice. 2020. Piccole finestre sul nonumano. Immagini per l'educazione ambientale nell'era della pandemia. *Visual Culture Studies* 2: 77–96.

Crutzen, Paul, and Eugene Stoermer. 2000. The Anthropocene. *Global Change Newsletter* 41: 17–18.

Daly, Natasha. 2020. Fake animal news abounds on social media as coronavirus upends life. *National Geographic*, March 20. https://www.nationalgeographic.com/animals/article/coronavirus-pandemic-fake-animal-viral-social-media-posts. Accessed 19 Nov 2022.

Diamond, Jared. 2005. *Collapse: How Societies Choose to Fail or Succeed*. New York, NY: Viking Penguin.

Di Paola, Marcello, and Gianfranco Pellegrino. 2018. La Terra reinventata. Etica dell'ambiente e Antropocene. *Semestrale Di Studi e Ricerche Di Geografia* 2: 85–102.

Erasmus, Desiderius. 1500–1536. *Adagia*. English Trans. Phillips, Margaret Mann, annotated by Mynors, Roger Aubrey B. 1989. *Collected Works of Erasmus*. Toronto-Buffalo-London: University of Toronto Press.

Gattei, Stefano, ed. 2013. *Natura senza dogmi. Per un approccio razionale al dibattito sull'ambiente*. Rome: Armando Editore.

Greimas, Algirdas J., and Joseph Courtés. 1979. *Sémiotique: Dictionnaire raisonné de la théorie du langage*. Paris: Hachette. English Trans. Crist, Larry, Daniel Patte, James Lee, Edward McMahon II, Gary Phillips, and Michael Rengstorf. 1982. *Semiotics and Language: An Analytical Language*. Bloomington, IN: Indiana University Press.

Grinevald, Jacques. 2007. *La Biosphère de l'Anthropocène: climat et pétrole, la double menace. Repères transdisciplinaires (1824–2007)*. Geneva: Georg/Editions Médecine & Hygiène.

Idone Cassone, Vincenzo, Bruno Surace, and Mattia Thibault. 2018. With a bang or with a whimper. In *I discorsi della fine. Catastrofi, disastri, apocalissi*, ed. by Vincenzo Idone Cassone, Bruno Surace and Mattia Thibault, 9–25. Rome: Aracne.

Jacobelli, Gian Piero. 2013. Tutti insieme rischiosamente. In *Natura senza dogmi. Per un approccio razionale al dibattito sull'ambiente*, ed. Stefano Gattei, 199–216. Rome: Armando Editore.

Kirchhoff, Thomas, and Vera Vincenzotti. 2014. A historical and systematic survey of European perceptions of wilderness. *Environmental Values* 23 (4): 443–464.

Le Roy, E. 1927. *L'exigence idealiste et le fait d'evolution*. Paris: Boivin & Cie.

Le Roy, E. 1928. *Les origines humaines et l'evolution de l'intelligence*. Paris: Boivin & Cie.

Lévi-Strauss, Claude. 1958. La structure des mythes. In *Anthropologie Structurale*, 227–255. Paris: Plon.

Lewandowsky, Stephan, Peter Jacobs, and Stuart Neil. 2022. Conspiracy theories made it harder for scientists to seek the truth. *Scientific American* 326 (3): 72–77.

Magdoff, Fred. 2013. The depletion of the world's natural resources: Is population the problem? *Monthly Review* 64 (8): 13–28.

Nisbet, Euan G. 1991. *Leaving Eden: To Protect and Manage the Earth.* Cambridge, UK: Cambridge University Press.

Painter, James, María Carmen. Erviti, Richard Fletcher, Candice Howarth, Silje Kristiansen, Bienvenido León, Alan Ouakrat, Adrien Russell, and Mike S. Schäfer. 2016. *Something Old, Something New: Digital Media and the Coverage of Climate Change.* Oxford: Reuters Institute for the Study of Journalism.

Pellegrino, Gianfranco. 2022. *L'Antropocene come età degli ibridi.* Presentation at "Seminari di Urbino 2022", Urbino, September 14.

Petitot, Jean. 1979. Hypothèse localiste et théorie des catastrophes. Note sur le Débat. In *Théories du langage, théories de l'apprentissage, le Débat Chomsky/Piaget*, ed. Massimo Piattelli-Palmarini, 516–524. Paris: Le Seuil.

Petitot, Jean. 1985. *Les catastrophes de la parole: De Roman Jakobson à René Thom.* Paris: Maloine.

Petitot, Jean, and René Thom. 1983. *Sémiotique et théorie des catastrophes—Actes semiotiques, Documents du Groupe de Recherches Sémio-linguistiques.* Paris: Institut National de la Langue Française.

Peverini, Paolo. 2020. Spazi "deserti" e vita sospesa ai tempi della pandemia. I droni e il ritorno alla Natura. In *Diario semiotico sul Coronavirus—E|C*, ed. Anna Maria Lorusso, Gianfranco Marrone and Stefano Jacoviello, 16–18. http://www.ec-aiss.it/index_d.php?recordID=1032. Accessed 16 Nov 2022.

Pietrabissa, Camilla. 2020. The eternal event. Urban void and image temporality from the renaissance to 2020. *Visual Culture Studies* 2: 15–33.

Revkin, Andrew C. 1992. *Global Warming: Understanding the Forecast.* New York, NY: American Museum of Natural History.

Ruthven, Malise. 2015. The apocalyptic social imaginary. In *Modernism, Christianity and Apocalypse (Studies in Religion and the Arts, 8)*, ed. Erik Tonning, Matthew Feldman and David Addyman, 354–383. Leiden: Brill.

Santana, Carlos. 2019. Waiting for the Anthropocene. *The British Journal for the Philosophy of Science* 70 (4): 1073–1096.

Shushakov, Egor. 2020. Генезис термина «ноосфера» и его использование П. Тейяром де Шарденом и В. И. Вернадским [Genesis of the term noosphere and its use by P. Teilhard de Chardin and V. I. Vernadsky]. *Вестник* 87 (87): 87–105.

Smolak Lozano, Emilia and Atsuho Nakayama. 2022. Análisis semántico del discurso sobre el cambio climático en social media. El enfoque de la opinión pública acerca de Greta Thunberg, incendios en Australia y COP25. *TECHNO REVIEW: International Technology, Science and Society Review/Revista Internacional de Tecnología, Ciencia y Sociedad* 14 (1): 1–20.

Stano, Simona. 2018. La catastrofe fra natura e cultura: riflessioni semiotiche sulla fine del mondo. In *I discorsi della fine. Catastrofi, disastri, apocalissi*, ed. Vincenzo Idone Cassone, Bruno Surace and Mattia Thibault, 29–37. Rome: Aracne.

Steffen, Will, Jacques Grinevald, Paul Crutzen, and John McNeill. 2011. The Anthropocene: A new epoch of geological time? *Philosophical Transactions: Mathematical, Physical and Engineering Sciences* 369 (1938): 842–867.

Teilhard de Chardin, Pierre. 1957 [1923]. L'hominisation. In *La Vision du passé*, 75–112. Paris: Éditions du Seuil.

Terrone, Enrico. 2018. Perturbazione e disintegrazione. La logica narrativa del genere catastrofico. In *I discorsi della fine. Catastrofi, disastri, apocalissi*, ed. Vincenzo Idone Cassone, Bruno Surace and Mattia Thibault, 85–98. Rome: Aracne.

Thom, René. 1973. Langage et catastrophes: Eléments pour une sémantique topologique. In *Dynamical Systems—Proceedings of a Symposium held at the University of Bahia, Salvador, Brasil*, ed. Maurício Matos Peixoto, 619–654. New York-London: Academic Press.

Thom, René. 1983. *Paraboles et catastrophes: Entretiens sur le mathématiques, la science et la philosophie.* Paris: Flammarion.

Tooze, Adam. 2020. We are living through the first economic crisis of the Anthropocene. *The Guardian*, May 07. https://www.theguardian.com/books/2020/may/07/we-are-living-through-the-first-economic-crisis-of-the-anthropocene. Accessed 11 November 2022.

Vernadsky, Vladimir. 1991. *Научная мысль как планетное явление*. Moscow: Наука. English Trans. Starostin, B. A. 1997. *Scientific Thought as a Planetary Phenomenon*. Moscow: Nongovernmental Ecological V.I. Vernadsky Foundation.

Waters, Frank. 1993. *Book of the Hopi*. New York, NY: Penguin Books.

Wright, Sidney Fowler. 1928. *Deluge: A Romance*. New York: Cosmopolitan Book Corporation.

## Filmography

Emmerich, Roland. 2004. *The Day After Tomorrow*. United States, 124'. Trailer: https://www.youtube.com/watch?v=Ku_IseK3xTc. Accessed 11 Dec 2022.

Emmerich, Roland. 2009. *2012*. United States, 158'. Trailer: https://www.youtube.com/watch?v=ce0N3TEcFw0. Accessed 11 Dec 2022

Feist, Felix E. 1933. *Deluge*. United States, 70'.

McKay, Adam. 2021. *Don't Look Up*. United States, 138'.

Stanton, Andrew. 2008. *WALL·E*. United States, 98'.

# Chapter 5
# "Uncooking" the Cooked: How to Eat Nature

**Abstract** The food domain is traversed by a number of mythologies concerning nature. This chapter explores some relevant case studies, ranging from "clean eating" to aphrodisiac substances, from genetic manipulations to the so-called organic paradigm, in order to identify the main meaning-making processes underlying them and the effects of meaning resulting from them. After an initial theoretical framing, some of the most important trends in contemporary foodspheres are analysed through a semiocultural approach, based on the discussion of significant examples and relevant literature in the considered field of studies. This offers crucial insights on the discursivisation of nature in food-related discourses and practices, pointing out a series of crucial inconsistencies and problematic aspects related to it.

## 5.1 From the Raw to the Cooked… and Back

In 1964 Claude Lévi-Strauss described cooking as a "technical activity" ensuring a transition between nature and culture: by cooking what is raw—he remarked—we transform it in a cultural product with symbolic meanings and values. Drawing on the correlation between specific conceptual pairs related to food (such as, precisely, "raw" *vs.* "cooked") and the corresponding oppositions on the semantic level (such as "nature" *vs.* "culture", or "unelaborate" *vs.* "elaborate"), he came to formulate the so-called "culinary triangle" (1964, 1965), where food is represented as a system located within a triangular semantic field, whose three vertexes correspond to the categories of the *raw*, the *cooked*, and the *rotten*. The raw represents the unmarked pole, while the other two vertexes are marked, although oppositely: the cooked is a cultural transformation of the raw, the rotten is its natural modification.[1]

[1] In fact, as Greimas (1983) effectively pointed out, it is essential to remark that these are not ontological realities. Raw food (e.g. a salad, a fruit salad, a beef tartare), in other terms, is not "natura", but means "nature"; similarly, cooked food (e.g. bread, a steak, a risotto), being more elaborate, means "culture"; and rotten food (e.g. particular types of cheese, beers, etc.), which is a natural transformation of raw materials—sometimes requiring particularly refined techniques—, stands between these two poles. The nature/culture opposition, in other words, necessarily relies on an effect of meaning resulting from different *perceptions* of the degree of elaboration of foods: the more they are elaborated, the more they are perceived as "cultural"; the less they are, the more

 
S. Stano, *Critique of Pure Nature*, Numanities - Arts and Humanities in Progress 26,
https://doi.org/10.1007/978-3-031-45075-4_5

As Lévi-Strauss highlighted, this model can also be used to distinguish various modes of cooking. *Roasting*, for instance, can be related to the raw, as it realises an unmediated conjunction between food and fire and does not provide uniformly cooked food. By contrast, both *smoking* and *boiling* entail a mediation (which is operated by air in the former, and by both water and the pot containing it and food in the latter), requiring a more evident transformation or "elaboration". While being double mediated, however, boiling results in foods that are far less durable than smoked products, which makes the scholar identify it with the rotten, while relating smoking to the cooked.[2]

Though certainly limited and unable to describe the complexity of human cooking systems and techniques (as Lévi-Strauss himself recognised in his writings[3]), this model is particularly interesting, as it highlights that "the cooking of a society is a language in which it unconsciously translates its structure" (Lévi-Strauss 1965, Engl. Trans. 2013: 35), representing on the level of expression certain systems of values, social relations, religious beliefs, and ideological investments.

Nevertheless, not all cultures and societies have attributed the same positive meanings and values to cooking processes. As Laudan (2004 [2000]) interestingly pointed out, for instance, a radical change occurred towards the middle of the 17th century. Before this time, cooking represented the very metaphor of all vital processes: the sun was thought "to cook" seeds into plants, fruits and grains; humans then "cooked" such substances into edible dishes; and finally their internal organs "cooked" the ingested food by digesting it, thus producing energy and vital substances, and returning wastes to the soil by expelling them—which allowed the cycle to start again. After 1650, this metaphor was replaced by that of "fermentation", which was associated with putrefaction, distillation, and the process of interaction of acids and salts. Although always starting with seeds and ending with the release of waste, the cosmic culinary cycle of that era involved a different approach to food itself, leading to a preference for products that fermented readily (such as oysters, anchovies, green vegetables and fruits) and hence did not need complicated preparation (i.e. "cooking") in the kitchen, with evident changes in both people's diet and perception of food.

Moreover, a marked "uncooking" trend, which seems to have gone well beyond the praise of foods not requiring elaborated cooking procedures (as it was in the socio-cultural contexts described by Laudan) to embrace a radical rejection and dysphorisation of the processes of "culturalisation" of food, can be observed in contemporary foodspheres. In order to better describe this phenomenon, some relevant case studies

---

they are perceived as "natural" (Marrone 2017: 24). However, as this chapter shows, this fact tends to be generally neglected or disregarded.

[2] As Lévi-Strauss noted, in fact, two levels can be distinguished: "the boundary between nature and culture … puts the roasted and the smoked on the side of nature, the boiled on the side of culture *as to means*; or, *as to results*, the smoked on the side of culture, the roasted and the boiled on the side of nature" (Lévi-Strauss 1965, Engl. Trans. 2013: 46). In other words, "it would seem as if the prolonged enjoyment of a cultural product involved, sometimes on the level of ritual and sometimes on that of myth, a corresponding concession in favour of nature: when the result is long-lasting, the means must be precarious, and vice versa" (Lévi-Strass 1968, Engl. Trans. 1978: 489).

[3] See in particular Lévi-Strauss (1965), cf. Buosi (2004), Stano (2015a).

will be described and analysed through a semiocultural approach in the following paragraphs: "clean eating" and the current "mania" for "natural" healthy diets; the lively debate on GMOs and organic foods; and the aphrodisiac potential traditionally ascribed to specific food products.

## 5.2 "Clean Eating", or the Taboo of Processed Food

The expression "clean eating" has become rather common in the last decades. Described by the Oxford Dictionary as "a diet consisting of unprocessed, unrefined, and nutrient-rich food" (Oxford Dictionary 2017: s.v. "clean eating"), it refers to a varied range of regimens, whose origin and evolution are not easy to define:

> It is very hard to pinpoint the exact moment when "clean eating" started, because it is not so much as a single diet as a portmanteau term that has borrowed ideas from numerous pre-existing diets: a bit of Paleo here, some Atkins there, with a few remnants of 1960s macrobiotics thrown in for good measure (Wilson 2017).

Nonetheless, it is possible to trace the emergence of clean eating as an autonomous food trend back to the beginning of the 21st century. More precisely, a first and moderate version of this trend is thought to have originated in 2007, with the publication of the book *The Eat-Clean Diet* by Tosca Reno, a Canadian fitness model. In it, the woman described how she lost a considerable amount of weight (34kg) and obtained various health benefits by eliminating "processed foods", and particularly white flour and refined sugar, from her diet. As a consequence, organic[4] fruits and vegetables, nuts, seeds, single-ingredient grains (such as quinoa, millet, oats, barley, brown rice, etc.), whole flour, raw cane sugar, and other "whole" foods—that is, products not having undergone substantial modification[5]—soon became the most praised ingredients of any diet wanting to be "clean". The attention, in other terms, was put on the "naturality" of food, intended in a differential sense, namely as the absence of processes "strip[ping food substances] of nutrients, bleach[ing] them, combin[ing]

---

[4] In fact, the consumption of organic products is often stressed as a crucial aspect of clean eating as a reassurance that they are as unprocessed and "natural" as possible. See *infra*, Sect. 5.3, for a dedicated section on the discursivisation of nature and naturality as related to the organic paradigm.

[5] According to the definition provided in the dictionary, a "whole food" is "a *natural* food and especially an unprocessed one" (Merriam-Webster 2022, my emphasis). Camille Adamiec also stressed this dimension in her analysis of the development of healthy food trends: "les produits raffinés ont perdu tous leurs nutriments, leurs vitamines, leurs oligoéléments soit l'ensemble de leurs vertus pour la santé. Si «les aliments, pour être parfaitement assimilés, doivent comporter l'ensemble équilibré dont les a dotés la nature» (Valnet 2010 [1985], p. 63), par contraste, les produits industriels sont au mieux des produits morts dans le sens où ils n'ont plus d'effets sur l'organisme, au pire des produits maléfiques puisqu'ils détruisent le corps et l'empoisonnent progressivement" (Adamiec 2013: 62) [*Refined products have lost all their nutrients, their vitamins, their trace elements, namely all of their health benefits. If 'to be perfectly assimilated, foods must include the balanced whole with which nature has endowed them'* (Valnet 2010 *[1985], p. 63), by contrast, industrial products are—at best—dead products, since they no longer have any effect on the body; or—at worst—harmful products, since they destroy the body and poison it gradually*, my translation].

chemicals, or other unnatural additives [with them]", and accused of making "the[ir] look, feel, and ... taste ... different from their *natural* form" (*Healthy Habits*,[6] my emphasis).

Just a few years later, another book, *Clean: The Revolutionary Program to Restore the Body's Natural Ability to Heal Itself*, published by Uruguayan former cardiologist Junger (2009), marked the birth of a more radical version of the clean eating trend, based on "detox" phases entailing the consumption of liquid meals and the total exclusion of caffeine, alcohol, dairy, eggs, sugar, all vegetables in the "nightshade family" (e.g. tomatoes, aubergines and so on) and red meat (which, according to Junger, creates an acidic "inner environment"), among several other foods. A number of other books similar to Junger's followed, also including worldwide best sellers such as Davis' *Wheat Belly*[7] (2011) and Perlmutter's *Grain Brain* (2013), which gave origin to a real "grain-phobia"[8] (see Stano 2018), actively contributing to increase the number of people who, in the last decades, have decided to avoid gluten despite not suffering from coeliac disease, just as other titles fomented mass obsessions for non-dairy milks and other free-from products and regimens. Moreover, a special praise for "nutrient-rich food[s] considered to be especially beneficial for health and well-being" (Oxford Dictionary 2022)—i.e. the so-called "superfoods"[9]—such as avocados, Goji berries or coconut oil, has progressively emerged. On the one hand, therefore, the "uncooking" process initiated by the first "clean" impulse has been confirmed and incredibly enhanced, making food regimens become real "conditions of being" (Stano 2021a) and making us become—to recall a popular expression introduced by anthropologist Marino Niola (2015)—*homini dietetici*, that is to say, obsessed by diet[10] and healthy food.[11] On the other hand, this has—quite paradoxically—resulted in the emergence of another way of thinking—another "culture", one might say—of food, which has made the nutritional dimension itself stand out as a real system of signification—and, more specifically, as a "system of classification" (cf. Douglas 1972)—of food, although in a reductionist manner and on the basis of a process of de-contextualisation, simplification and exaggeration of its basic units, i.e. "nutrients".[12]

---

[6] https://www.healthyhabithhi.com/blog/wholefood-vs-processed (Accessed: 4 December 2022).

[7] Reprinted and expanded in 2014 as *Wheat Belly Total Health* (Davis 2014).

[8] And even a real conspiracy theory, see Stano (2016, 2021c).

[9] Despite its ubiquity in the mass and especially new media discourses, there is no official or legal definition of the term "superfood" (EUFIC 2012). While increasing efforts have been made to regulate the use of nutrition and health claims in food marketing to ensure that food's labelling, presentation and advertising is clear and accurate (consider, for instance, the regulation (EC) No 1924/2006 issued by the European Union), several new, generally exotic products have been increasingly celebrated as "superfoods", becoming largely diffused in Western markets, even though the exceptional nutrient content ascribed to them is not supported by scientific evidence.

[10] Niola (2015: Chap. 2) provocatively argues that if once man made his diet, now it is the diet that "makes" man.

[11] Refer to the discussion on orthorexia (see *infra*).

[12] Gyorgy Scrinis criticised this phenomenon as the *ideology of nutritionism*: "Nutrition scientists, dieticians, and public health authorities—the nutrition industry, for short—have implicitly or

As I highlighted in Stano (2021c), the role of social media in the spread of such trends and regimens must be noted. In fact, the books generally represent only one of several nodes in a dense intertextual network, which functions like a real "echo chamber" tending to generate homogeneous clusters (Del Vicario et al. 2016) and favouring group polarisation.[13] This also triggers an effect of "emotional contagion", which is strictly connected with the so-called "post-truth" phenomenon and, more generally, a progressive fading of the separation between fact and fiction (as a further accentuation of the dynamics noted by Habermas 1971 [1962], Engl. Trans. 1991: 170).[14] Thus, more or less founded theories and dietary prescriptions have been able to leave the pages of books such as *Wheat Belly* or *Grain Brain* to delve into the pervasive networks of the digital media, thanks to which they have spread globally, becoming major trends in current foodspheres. Therefore, it should come with no surprise that, as Wilson (2017) remarked, almost all of the most influential "clean eaters" are bloggers and/or Instagrammers. The case of Jordan Younger, better known as "The Blonde Vegan",[15] is emblematic in this sense: a "wellness" blogger with almost a million followers on Instagram and no qualifications at all as a nutritionist, she sold more than 40,000 digital copies of her 25-$ five-day "Cleanse Program",[16] based on an all-raw, gluten-free, plant-based almost entirely liquid diet—which then proved extremely harmful and she herself had to abandon.[17]

On the other hand, criticisms towards the phenomenon also appeared and spread widely through several media, just a few years after it affirmed itself as a major food trend on a global scale. In 2015, for instance, the famous food writer and television cook Nigella Lawson expressed her "disgust"[18] for clean eating, claiming that "food

---

explicitly encouraged us to think about foods in terms of their nutrient composition, to make the connection between particular nutrients and bodily health, and to construct "nutritionally balanced" diets on this basis. … This focus on nutrients has come to dominate, to undermine, and to replace other ways of engaging with food and of contextualizing the relationship between food and the body (Scrinis 2008: 39; see also Scrinis 2013). Pollan (2008) also insisted on the problems brought about by this new way of looking at food, focusing in particular on the case of margarine.

[13] In fact, as illustrated by Quattrociocchi and Vicini (2016), users involved in a community are more likely to focus on specific topics and viewpoints, "isolating themselves" from their surroundings, that is to say, from alternative topics and viewpoints.

[14] In fact, clean eating has been increasingly described as a "post-truth cult" (Wilson 2017), namely "part of a post-truth culture, whose adherents are impervious, or even hostile, to facts and experts" (*Ibidem*) and hence favour personal experience over scientific data (see Fivian and Wood 2019).

[15] Now rebranded as "The Balanced Blonde", after a radical change in her habits and image (see *infra*).

[16] https://thebalancedblonde.gumroad.com/l/theblondevegancleanse?layout=profile (Accessed: 30 November 2022).

[17] As a consequence of her unbalanced diet, Younger's hair started falling out, her skin turned orange and she lost her period. A huge debate was raised by her 2014 post "Why I'm Transitioning Away from Veganism" (https://thebalancedblonde.com/2014/06/23/why-im-transitioning-away-from-veganism/), not without ideological distortions and manipulations of her message. What is relevant to consider in relation to the issues described in this chapter is her recognition of her "disordered eating habits" and the unhealthy regimen of food restrictions sha had fallen into.

[18] https://www.bbc.co.uk/sounds/play/p034s2dx (Accessed: 30 November 2022).

is not dirty" (Lawson 2015: II), and strongly opposing the movement and its "puritanism". In 2016, the journalist Ruby Tandoh published an article in VICE magazine, describing clean eating as "an incitement to eating disorders" (Tandoh 2016) and the following year Dr Giles Yeo, a geneticist at the University of Cambridge, also reinforced this idea through an episode of the BBC's *Horizon* series entitled "Clean Eating: The Dirty Truth"[19] (Quinn 2017), which saw him interviewing some of the most celebrated representatives of the movement (including Ella Woodward, aka Ella Mills, creator of the food blog and brand *Deliciously Ella* and author of various cookbooks[20]; Dr William Davis, author of the above-mentioned *Wheat Belly* and a leading promoter of grain-free diets; and Dr Robert O. Young, the "godfather" of the Alkaline diet, who inspired Natasha Corrett and Vicki Edgson's *Honestly Healthy* (2012)) to highlight the absence of scientific evidence and the "troubling narratives" underlying most of their claims and communicative channels and products.

As the negative press for clean eating intensified, many of its representatives tried to rebrand themselves, distancing from the word "clean" and the absolutism and extremeness attributed on it. Ella Mills, for instance, declared in the end of the BBC documentary hosted by Yeo that she no longer recognised herself into the movement: "When I first read the term it meant natural, unprocessed. … Now it doesn't mean that at all. It means diet. It means fad" (*Clean Eating: The Dirty Truth,* 57′ 00"–57′ 07"). Similarly, the popular Hemsley sisters refused to take part in the film, releasing a statement distancing them from "absolutes", as well as from "universal" regimens.[21]

Yet, as a 2017 article published on *The Guardian* interestingly remarked, "The oh-so-Instagrammable food movement has been thoroughly debunked—but it shows no signs of going away" (Wilson 2017). In fact, most of its representatives continue to spread recipes and advice on "healthy" lifestyles and practices, sometimes in more moderate tones, sometimes with no changes at all. And it has evidently changed the products available to all of us, as well as the way they are spoken and thought of. Consider, for instance, the emblematic case of avocados, which have come to outsell oranges in the UK (cf. *Ibidem*) and have imposed themselves on market and supermarket shelves all over the world.

Powered by social media, clean eating, in all its facets, has become more popular than any previous trend or school of modern nutritional advice, fomenting the crucial problem of "orthorexia", that is to say, a diffused obsession for "pure", "healthy" and "perfect" food (McCartney 2016; cf. Bratman 1997; Bratman and Knight 2001; Nicolosi 2006–2007, 2007; Wynne et al. 2021). Moreover, as related to the specific topics discussed in this book, it has resulted in an ambivalent, and highly problematic position. In fact, as it was highlighted above and the commentary by Mills in the BBC

---

[19] https://archive.org/details/BBCHorizonCleanEatingTheDirtyTruthFullDocumentary (Accessed: 29 November 2022).

[20] E.g. *Deliciously Ella* (Woodward 2015), *Deliciously Ella Every Day* (Woodward 2016), *Deliciously Ella With Friends* (Woodward 2017).

[21] "We asked the Hemsley sisters, who cut out all grains, to take part in this film, and they refused. In a statement, they told us, 'Grains are already abundant in the modern diet so our recipes celebrate other ingredients'. They told us they don't believe in absolutes and no one way of eating suits everyone" (*Clean Eating: The Dirty Truth,* 22' 41"–22' 54").

documentary also reflects, the trend originated precisely in the name of "nature", that is to say, as an attempt of "cleansing" food from the excessive culturalisation ascribed to contemporary food systems. However, as the examples discussed above show, it has resulted in an even more "culturalised" set of regimens, where food is mainly reduced to a set of nutrients, such as acids (e.g. the Alkaline diet) or proteins (e.g. gluten-free regimens), etc., in a reductive and often simplistic approach that—if we maintain the dichotomy on which the very definition of clean eating is based—is not "natural" at all.

## 5.3 GMOs *Versus* Organic Food: A Never-Ending Debate

Another crucial issue in the debate concerning food and nature is the opposition between genetically modified foods (GM foods, or GMOs), on the one hand, and organic food, on the other hand. Though scientific consensus on the safety of the former has increased in recent years, most people still perceive them as a major threat for health, as they are mainly seen as "unnatural". By contrast, the latter are more praised than ever precisely for the opposite reason. As I showed in Stano (2021a), while certainly linked to material factors, such trends are also evidently dependent on the diffused mythology of "nature" marking contemporary foodspheres.

In fact, GMOs are defined as "organisms in which the genetic material (DNA) has been altered in a way that does not occur *naturally* by mating and/or *natural* recombination …" (WHO 2014, my emphasis). Largely shared and adopted by a number of institutional entities,[22] with only slight variations, such a description is highly problematic precisely because of its reference to "nature":

> It relies on manifestly arbitrary and not at all clear criteria for defining the "naturalness" of such techniques. According to the European legislation, for instance, some practices, such as the recombination of nucleic acids, cell fusion, or the injection of external heritable material into one organism, are considered forms of genetic modification, while others, such as the so-called mutagenesis or some methods of cell fusion not involving nucleic acids recombination, are not (cf. Breassanini and Mautino 2015; Stano 2021b). This originates a series of borderline cases, as well as huge differences in legislations, with the US law HR 933 (2013) preventing federal judges from introducing any ban on the sale of GMOs (not without oppositions and bottom-up initiatives increasingly manifesting discontent towards such a regulation) versus Europe adopting a "precautionary principle" that bans any recognised genetic modification of food (Stano 2021a: 154).

Similarly, the organic paradigm, which is celebrated for "respecting *nature*" (Rodale 2010), also entails huge indeterminateness, with organic products mainly

[22] Consider, for instance, the European Directive 2001/18/CE (https://eur-lex.europa.eu/legal-content/EN/TXT/?uri=CELEX%3A32001L0018), the American National Bioengineered Food Disclosure Standard (2018, https://www.federalregister.gov/documents/2018/12/21/2018-27283/national-bioengineered-food-disclosure-standard), and the popular the NON-GMO Project (https://www.nongmoproject.org).

defined in differential terms, that is to say, in opposition to "anything that is non-synthetic" (as well reflected by the description provided by the Food and Agriculture Organization of the United Nations, cf. FAO 1999).

This same configuration is reflected by most collective representations of both organic products and GM foods: while the former generally celebrate the absence, and even the deliberate rejection of culture (as demonstrated by the "Organic. Naturally Different" campaign launched in 2011 by the Organic Centre Wales (OCW), where fruits, vegetables and cows show off their pride for not "hav[ing any] work done", which makes them "*naturally* different"[23]), the latter attack precisely the "monstrosity" and the "subversion" of the natural order operated by the genetic (i.e. cultural) modification of food (with oranges-kiwis, apples-watermelons and other visual paradoxes recalling the figure of Frankenstein, along with verbal expressions such "Frankenfoods", "zombie foods", etc., and popular personalities and institutions reproaching food biotechnology and science for going "against the God's will"[24] and warning the consumers about the consequences of such cultural destabilisations on nature itself). A further dysphorisation of this type of "culturalisation" of food can also be noted on the level of veridictory relations (cf. Greimas 1966): as I illustrated in the essay "Food, Health and the Body: A Biosemiotic Approach to Contemporary Eating Habits" (Stano 2021b), while the lack of processes of alteration in organic products creates an effect of "truth", making such products *appear as they are*, transgenic foods rather oscillate between "secret" (*they do not seem what they are*) and "lie/illusion" (*they are not what they seem*).

While reminding the importance of reflecting on the possible risks deriving from such processes of manipulation of food, the dynamics recalled above also highlight an excessive simplification of the opposition between a supposed "respect for nature", on the one hand, and its subversion by means of its "extreme culturalisation", on the other hand. The very idea of nature, moreover, emerges, once again, as extremely unclear and variable, which makes discourses on this topic, also including regulatory documents and international agreements, exposed to great indeterminateness and ideological investments.

To conclude, it is crucial to highlight that, such dynamics, and the effects of meaning deriving from them, are not a peculiarity of the debate on GMOs, but extend to several other aspects concerning the technological innovation of food systems and products. A recent Italian campaign[25] against the production, use and sale of the so-called "cell-based foods"[26] is particularly emblematic in this sense. A poster

[23] For a detailed description and analysis, see Stano 2021b: 52–53. On the communication of organic products and naturality, see also Marrone (2011) and Ventura Bordenca (2012).

[24] As said by Pope John Paul II, quoted by Lyman (2000).

[25] Available at https://www.coldiretti.it/economia/consumi-gia-200mila-firme-contro-cibo-in-provetta (Accessed: 30 November 2022).

[26] Whose variable denomination (ranging from expressions such as "clean", "animal-free", "slaughter-free" products to "cultivated", "cultured", "synthetic", "craft", "artificial" foods") is also particularly interesting as related to the mythologies of naturality in the food domain (see in particular Buscemi 2015; Szejda et al. 2019; Bertero et al. 2023).

released in November 2022 to support the campaign relied precisely on the above-mentioned idea of "nature": the subdivision of the page into two main columns was used to emphasise the opposition between two main categories, which the verbal code identified as "CIBO NATURALE" [*NATURAL FOOD*] versus (with the usual formula "VS" just in the middle) "CIBO SINTETICO" [*SYNTHETIC FOOD*], directly addressing the reader to ask him/her "Da che parte stai?" [*What side are you on?*, my translation]. At the figurative level, the first category found expression in pictures of grazing cows, cured meats on a table, and some farmers taking care of their fields and/or showing off their products. On the contrary, the reference to "synthetic" foods was figurativised through sketches[27] of scientists wearing nuclear suits and managing test tubes emitting smoke, just above a rather unusual machine and the drawing of a plumbing releasing toxic substances into the environment (as the leafless tree besides it suggested). A skull and crossbones symbol was also used to recall the ideas of death and danger, as it is usual for poisonous chemicals. A list of pros (related to "natural food") and cons (related to "synthetic food") followed, with an interesting chromatic configuration recalling safety and nature (i.e. the use of green and light blue) on one side, and danger and alarm (i.e. red and dark grey) on the other side. Such a valorisation was also stressed in a statement that was released in support of the campaign, and especially in its final sentences: "Le bugie sul cibo in provetta confermano che c'è una precisa strategia delle multinazionali che con abili operazioni di marketing puntano a modificare stili alimentari *naturali* fondati sulla *qualità* e la *tradizione*. … Siamo pronti a dare battaglia poiché quello del *cibo Frankenstein* è un futuro da cui non ci faremo mangiare" [*All the lies about synthetic food confirm that multinational corporations have a particular strategy to modify* natural *food styles based on* quality *and* tradition *through devious marketing moves. … We are ready to fight [them] because* Frankenstein food *is not a future scenario we will let ourselves be eaten by*, my translation and emphasis]. Nature was thus explicitly related to a static tradition (to which quality was directly connected), that is to say, to a "crystallised" idyllic past to be preserved and protected from any kind of innovation—which, on the contrary, was portrayed, exactly as observed for the collective imaginary of GMOs, as monstruous and disrespectful of human limits,[28] as the reference to Mary Shelley's famous character (see Shelley 1818) remarked. But is not tradition itself the result of a series of cultural operations of selection of techniques, products, and practices? And are not the foods praised as "natural" on this poster themselves the products of cultural processes such as cultivating fields, cropping, farming, and even slaughtering (i.e. all practices whose impact on nature, intended both as the environment and the very life of animals, has been increasingly shown off and questioned)? This is not the place to discuss the complex political and ethical aspects related to such issues, but it is crucial to highlight how their presence

[27] Whose inclusion is particularly relevant also on the plastic level: while the pictures used for "natural foods" suggest an effect of reality, in fact, sketches and drawings rather mark a detachment from it.

[28] In fact, as Niola (2012) and Ortoleva (2019) observed, the representation of science and technology as manifestations of a dysphoric culture daring to go beyond the limits of humanity, eventually succumbing to its own creations, is a widespread modern mythology.

confirm the ideological operations that are at work any time "nature" is called upon to contrast innovation and change in the food domain.

## 5.4 The Mediterranean Diet: A "Natural" Regimen?

As mentioned above, it is not uncommon, in food discourses, to find a direct association between "nature" and "tradition", with the latter being interpreted as a return to a supposed original and idyllic past, crystallised in "authentic" recipes, "typical" habits, and so on and so forth.

An emblematic case in this respect is the Mediterranean diet, whose peculiar origin and evolution I illustrated in detail inStano 2015b.[29] It suffices here to remind that it originated from the scientific field, in the wake of medical research on heart and vascular diseases, being defined by physiologist Ancel Keys in terms of a dietary pattern characterised by a high intake of bread, legumes, vegetables, fruits and fats rich in unsaturated fatty acids (e.g. olive oil), a moderate intake of fish, and a low intake of dairy and meat (Keys 1980). Subsequently, it extended beyond the simple definition of healthy rules regulating nutrition to embrace the social and cultural implications of a specific "lifestyle". In fact, in 2010 the Mediterranean diet was included in the UNESCO Representative List of the Intangible Cultural Heritage of Humanity, and it was described precisely as a set of cultural practices (cf. Tylor 1871):

> The Mediterranean diet involves a set of skills, knowledge, rituals, symbols and traditions concerning crops, harvesting, fishing, animal husbandry, conservation, processing, cooking, and particularly the sharing and consumption of food. Eating together is the foundation of the cultural identity and continuity of communities throughout the Mediterranean basin. The Mediterranean diet emphasizes values of hospitality, neighbourliness, intercultural dialogue and creativity and plays a vital role in cultural spaces, festivals and celebrations, bringing together people of all ages, conditions and social classes (UNESCO 2013).

However, an anchorage to "nature" seems to be an unavoidable element of the collective representations of the Mediterranean diet, even as related to its inclusion in the UNESCO list. I will consider in the following some relevant case studies in order to explore how this occurs, and what are its effects in terms of meaning-making processes.

In (2013) the Italian Ministry of Agricultural, Food and Forestry Policies decided to celebrate the confirmation of the inclusion of the Mediterranean diet in the UNESCO Representative List of the Intangible Cultural Heritage of Humanity with a TV commercial directed by Nicola Paparusso.[30] Since the first monochrome scenes, the isotopy of "rural tradition" emerges: in the first sequence, a cart pulled by a horse,

---

[29] This paragraph builds on the research presented in the mentioned paper, selecting and further expanding the issues and examples that are particularly relevant to the purposes of this book.

[30] Available at https://www.youtube.com/watch?v=NRI3Q3ESKN8 (Accessed: 30 November 2022).

which brings a young woman, her husband and their son back home through a sunny street in the countryside, dominates the screen. A break is marked by the arrival of a military car (which, by contrast, represents a symbol of modernity), driven by two American soldiers. When these meet a young boy in a field of olive trees (i.e. another figure commonly used to recall the Mediterranean area), they ask him—in a broken Italian—where they can eat macaroni and wine. The young boy answers with the stereotyped paraverbal sign expressing his inability to understand them, suggesting a substantial incompatibility between the two semiospheres (cf. Lotman 1984) that these characters represent. This is when the animated rhythm of Glenn Miller's *In the Mood*, which opened the commercial, gives the way to Dean Martin's *Mambo Italiano*, whose first verses, sung by a choir, recall the acoustics typical of folk songs and also explicitly evoke the Italian context through their lyrics ("A boy went back to Napoli because he missed the scenery / The native dances and the charming songs"). In the meanwhile, some rapid sequences show a young and an elder woman dancing, then a plough tilling the soil, and finally the previously introduced young couple joyfully running through a field of spikes. Then the camera moves to some fruits and vegetables on a big table, and ultimately to the sensual dance of another young couple, whose harmonious movements reinterpret traditional food rites (such as setting the table or pressing the grapes by feet, as it was usual in Italian rural societies—but is not common at all today) with a ludic tone. Suddenly, the American soldier reappears on the screen, moving closer to the young lady dancing on the grapes and offering her an apple. She winks at him and takes the fruit, but pushes him away, just a second before another young bare-chested boy—whose outfit reveals his rural, and therefore local identity—, arrives and takes her off the barrel to continue dancing with her. This is when the second, evidently shorter, part of the commercial begins, with a succession of polychromatic figures referring to different cultural backgrounds (e.g. some ideograms evoking Asia, the tour Eiffel recalling France, and the Statue of Liberty representing the United States) opening the way to the very final sequence, set in a restaurant, where the same child shown in the opening brings some "maccheroni" to the American soldier, who is now dressing civil clothes. The screen then turns black to host the white message: "DIETA MEDITERRANEA AMBASCIATRICE DELLA SALUTE NEL MONDO" [*THE MEDITERRANEAN DIET: AMBASSADOR OF HEALTH IN THE WORLD*, my translation]. Finally, the emblem of Italy and the name of the Italian Ministry of Agricultural, Food and Forestry Policies appear on the screen, together with the Italian flag, to legitimise the entire message.

"Here" and "there", local and global, past and present, at first radically opposed (so as to create incomprehension and incommunicability), find in the end their point of contact in the Mediterranean diet—which is presented more as a "lifestyle" than as a set of healthy food choices. However, it remains evidently related to the idea of a sort of "innate", "natural" feature characterising those who live in the Mediterranean area. This is particularly evident in the scene involving the young woman pressing the grapes, especially as a result of the opposition between the two men who approach her: the American soldier and the local (i.e. Mediterranean) man. Both have toned and evidently healthy bodies (recalling the idea of "form", as described by Boutaud 2013), but while the former owes it to his particular *doing* (i.e. the military regime, as his uniform remarks), the latter seems to "naturally" possess it (thus rather recalling a specific *being*), as his "innate" lightness and easiness of movement (which the American soldier lacks) suggest. This highlights a *positional regime* (in the terms introduced by Ferraro 1998[31]), according to which participating in the harmonious dance of the Mediterranean diet requires belonging to the represented rural "here" (whose attachment to the past, as highlighted above, is particularly stressed), while no access is given to "others" coming from any "there" (as exemplified by the young lady's rejection of the American soldier, or the lack of communication between him and the Mediterranean child in the beginning of the commercial). Yet the end of the advertisement somehow introduces a "glimpse" of inclusion, partially recalling a *multiperspective regime* by means of specific figures (i.e. the ideograms, the Eiffel Tower and the Statue of Liberty, as well as the dialogue between the involved characters) and plastic formants (i.e. polychromy).

Nevertheless, such an opening is not always euphoricised. In fact, sometimes, the emphasis is rather put on the risks it entails. This happens, for instance, in the commercial[32] released in 2015 by Italy's national public broadcasting company RAI on the occasion of EXPO (2015) "Feeding the Planet, Energy for Life".[33] The initial part of the advertisement features most of the elements identified in the previous example: while the rural tradition is recalled (especially on a figurative level, through images such as a plough tilling the soil, an aged hand cutting a bunch of grapes and another one picking up a tomato), the voice over recalls the origin of the Mediterranean diet and its relation to a particular "here" (e.g. "nella Campania povera del dopoguerra" [*in poor post-war Campania*], "La stessa cosa la osserva nel resto del nostro Sud e in Grecia" [*the same fact was observed in other southern Italian regions and in Greece*, my translation]). A specific mention to the 2010 inclusion in the intangible heritage of humanity by the United Nations is also included, as the camera moves from the rural background to a metropolitan reality, showing the United Nations headquarter and finally a restaurant. This is when the relaxed rhythm characterising rural places and subjects suffers a sharp acceleration, leading to the frenetic movements of some waiters (accentuated through a super speed effect), who are in fact barely visible. The voice over then introduces the crucial issue animating the commercial: "La dieta mediterranea, insomma, si è globalizzata, ma la seguono soprattutto quei salutisti

[31] Building on the Greimassian opposition between the subjective level, which is related to the perspective of the subject, and the objective level, which rather stresses socially and intersubjectively recognised values, as well as on the contrast between a relative and an absolute dimension, Ferraro (1998) identified four "discursive regimes". In the *causal regime*, the emphasis is put on objective facts, and what one does defines what one is. In the *positional regime*, instead, what one is defines what one does; this is the realm of tradition and socially recognised roles. The third discursive regime is the *perspective regime*, which is based on the refusal of any external definition, in a subjective perspective. Finally, the *multiperspective regime* consists in grasping other people's desires and breaking into their narrative programs as a value-object, combining different perspectives.

[32] Available at https://www.youtube.com/watch?v=l9_lDREniYU (Accessed: 30 November 2022).

[33] I.e. a world's fair dedicated to food, with respect to different aspects, from sustainability of agricultural practices to haute cuisine, from culinary traditions to eating disorders, etc.

che si possono permettere vegetali freschi e pesce, piatti ben cucinati e ore di sport alla settimana" [*In sum, the Mediterranean diet has been globalised. But it is now mainly followed by health-conscious people who can afford fresh vegetables and fish, well cooked meals, and hours of sport a week*, my translation]. The opening of the Mediterranean diet to other places and people, in other terms, is presented as a loss for the places where it originated. In fact, after showing images of the mentioned health-conscious people, the commercial provocatively alludes to its disappearing from the previously described "here" ("Lì dove è nata, invece, non esiste quasi più" [*By contrast, where it was born, it has almost disappeared*, my translation]), while on the screen the hand of a child (recalling new generations) quickly "steals" an apple (here used as a symbol of health) from an elder hand (symbolising tradition). An infographic therefore comes in to illustrate better this issue, together with the voice over: "I consumi di ortaggi e frutta scendono anno dopo anno, e aumentano quelli dei cibi spazzatura. Oggi, proprio in Campania, quasi una persona su due è sovrappeso, e una su dieci è obesa" [*The consumption of vegetables and fruits decreases year after year, while that of junk foods increases. Today, precisely in Campania, almost half of the population is overweight, and one person in ten is obese*, my translation]. The voice over also informs that the problem is not limited to Campania, but concerns the wider Mediterranean area: "E il resto del Sud, come la Grecia, non sta molto meglio" [*And the other southern regions, as well as Greece, are not better*", my translation]. Hence the commercial ends with a provocative question: "La dieta mediterranea tornerà mai a casa?" [*Will the Mediterranean diet ever come back home?*, my translation]. While in rural past societies the *positional* regime described above was possible, in other terms, in globalised present societies the "Mediterranean way" seems to be no longer enabled by geographical or cultural belonging (i.e. a "being" that is natural to human beings), but only by people's will and actions (i.e. a "doing" characterising only a few individuals—the health-conscious ones—who play sports and do not succumb to junk foods), in a *causal* regime (cf. Ferraro 1998).

This commercial and its main message were strongly criticised by several organisations for promoting of the Mediterranean diet. The online magazine La Dieta Mediterranea.EU,[34] for instance, responded to it by digitally modifying its last frame and, more specifically, by adding the writing "Veramente… non si è mai mossa di casa" [*Actually… it never went away*, my translation] as a response to its final provocation, therefore re-attesting a *positional* regime that connects the Mediterranean diet with traditions and skills that are considered "con-natural" to the Mediterranean area. A series of similar initiatives and messages also followed.

In this sense, as Ventura Bordenca (2015) remarked, the Mediterranean diet can be defined as a "non-diet": while common regimens rely on regulating systems requiring the subject to strictly follow a number of peremptory prescriptions and procedures (a specific *doing*), it is considered a safe nutritional model "by definition", which does not require to follow any particular rule to be healthy and beautiful (unless the interference of globalisation comes in, according to RAI's commercial). In fact, as Boutaud (2013) observed, common diets can be distinguished into diets relying

[34] http://www.ladietamediterranea.eu (Accessed: 30 October 2022).

on the deontic modality of prohibition (*having not to do*) and diets relying on the deontic modality of prescription (*having to do*). In the case of the Mediterranean diet, such a "doing" is replaced by a specific "being", which is presented as an innate characteristic, only acquirable by "natural inheritance". Accordingly, the considered advertisements show that openness can only concern "a taste of it" (as in the end of the commercial directed by Paparusso) and might even have dangerous effects (as RAI's commercial warns).

While still largely shared, such representations and valorisations of the Mediterranean diet reveal a series of remarkable inconsistencies. Firs of all, it must be noted that the very idea of the "Mediterranean" is not clear at all, since it refers to a "common ancient background on which, from Goethe onwards, the image of modernity has been created"[35] (Niola 2014: 9, my translation), namely a stereotyped image that, although having been established elsewhere, has increasingly influenced the local representations of the "Mediterranean identity". Moreover, even though the lifestyle described by the United Nations has been historically shared by several groups living in the Mediterranean area, it is not possible to deny the significant differences among the various Mediterranean diets (in the plural form), which are in fact very diverse. This is reflected by the extreme variability of the geographical contexts to which different definitions and representations of the Mediterranean diet make reference: while the commercials put the emphasis on Southern Italy (both of them) and Greece (RAI's advertisement), in other cases different areas are also involved. The definition provided by the UN, for instance, first included Italy, Greece, Spain and Morocco, to which Portugal, Croatia and Cyprus were added in 2013. By contrast, the Bio-Med (*Bio-Mediterraneo—Salute, bellezza e armonia*) Cluster at EXPO 2015 encompassed Albania, Algeria, Egypt, Lebanon, Malta, Montenegro, San Marino, Serbia, and Tunisia, thus evidently differing from the UNESCO list.

Finally, it should be remarked that not only the geographical, but also the temporal characterisation of the Mediterranean diet and its references to nature seem highly problematic. As observed above, both the figurative level (e.g. elderly characters and old-style machines) and the plastic dimension (e.g. monochromy or the adoption of slow motion) recall an a-historic and archetypal conception of rural life, whose naturality finds expression in the idea of an immutable, crystallised "tradition" (*vs.* the threats brought about by the evolution of lifestyles). In fact, the Mediterranean diet was initially defined as "a nutritional model remained constant over time and space" (UNESCO 2012). As the United Nations themselves amended in 2013, however, it should be rather conceived as a "heritage *re*-created day after day" (UNESCO 2013, my emphasis), thus not refusing the idea of a "traditional" Mediterranean diet, but also admitting that, as any other tradition, it is constantly remodelled—and, to some extent even re-"invented" (cf. Hobsbawm and Ranger 1983)—, day after day. Yet, as the analysed examples remark, the collective imaginary of the Mediterranean diet still seems to remain anchored to a static view unable to adequately recognise and promote this transformative, and unavoidably cultural, character.

---

[35] "Sfondo antico su cui, da Goethe in poi, è stata disegnata l'immagine della modernità".

## 5.5 Aphrodisiac Foods, or the Seductive Potential of "Nature"

Countless online posts and magazine articles relentlessly promise their readers to reveal them how to "improve sexual life"[36] (*Gentleman's Journal*, undated), "totally up [their] sex drive"[37] (*Cosmopolitan* 2021), "boost [their] libido"[38] (*Healthline* 2020), and "arous[e] sexual instincts, brin[g] desire, or increas[e] sexual pleasure or performance" (*Ibidem*) by means of food. This is the realm of so-called "aphrodisiac foods".[39]

But how can exactly a product stimulate sexual desire? While, on the material level, there are in fact substances in some foods, such as testosterone, papaverine or phenylethylamine, which seem to have a certain effect on particular bodily functions related to sexual organs, there is no clear explanation or scientific evidence of their effective capacity to enhance the libido or improve sexual performance.

Let me briefly consider some of the most relevant examples in the history of aphrodisiac foods (as effectively described by Danieli 2019, among others) to understand how exactly they have been conceived over time.

In Ancient Egypt, aphrodisiacs included mainly vegetable foods (cf. Martinis 1999), such as anti-inflammatory and purifying plants like lettuce (which was also considered sacred to the god of fertility Min for its phallic shape) and antioxidant fruits like pomegranates (whose abundant grains allegorically referred to fertility). A special praise for spices as enhancers of sexual functions also arose in that era, with special recipes indicating how to combine wine with coriander, ginger and radishes mixed with honey. As it can be noted, it was not merely a question of material factors, but also of symbolic investments.

The symbolic dimension then became even more important in the Greek culture, as a particular aphrodisiac potential was projected on seafoods such as oysters, shellfish and caviar—which are still considered powerful stimulators of sexual activity—, by virtue of classical mythical references.[40]

The Romans also had special elixirs to increase sexual performance, which were called *amatorial pocula*, and were made by combining vegetable ingredients with toad hearts, other animal organs, and even human remains and secretions. Once again, the symbolic dimension is fundamental for understanding such "recipes": a typical

---

[36] https://www.thegentlemansjournal.com/10-best-aphrodisiac-foods/ (Accessed: 4 December 2022).

[37] https://www.cosmopolitan.com/sex-love/advice/g1022/aphrodisiac-foods-0509/ (Accessed: 4 December 2022).

[38] https://www.healthline.com/nutrition/aphrodisiac-foods (Accessed: 4 December 2022).

[39] From Ancient Greek *ἀφροδισιακός* (*aphrodisiakós*), derived from *Ἀφροδίτη* ("Aphrodite"), the term *aphrodisiac*, referred to food, in fact etymologically refers to substances "stimulating sexual desire" (Oxford Dictionary 2022).

[40] More specifically, according to Hesiod's *Theogony*, the goddess Aphrodite was generated precisely from sea foam, made fertile by Uranus' testicles, which had been hurled into the Aegean waters by Kronos (Ieranò 2019).

ingredient of these elixirs, for instance, was hanged men's fat, not because of any real effect connected to its intake, but rather because following the pressure of the noose on the cerebellum, hanged men were likely to die with a surprising erection, which used to remain post-mortem. The mandrake itself ended up being considered an aphrodisiac as a result of this, since it often sprouted under the scaffolds where hanging occurred (*Ibid*.: 160).

While the progressive diffusion of Christianity brought the sexual act back into the sphere of fecund marital chastity, dysphoricising sexual pleasure and exciting substances (see in particular Montanari 2016), the search for aphrodisiac recipes increased with Humanism and the Renaissance, which recovered and further expanded classical heritage in "food eroticism". This led, for instance, to Leonardo Da Vinci's famous recipe of the "acquarosa" (1517), described by the artist and scientist himself as a way to "excite Venus" (*eccitar Venere*) and "erect penis" (*dirizzar la verga*) by virtue of its vasodilator effect and other processes activated by its ingredients (see Danieli 2019: 162–163). Moreover, countless formulae (see in particular Camporesi 1989) enhancing the symbolic meaning of testicles of bulls, donkeys and other animals believed to be lewd "by nature" since ancient times (see Angela 2012) appeared, together with recipes based on other "extra-ordinary" foods (including "exotic" products such as tomatoes, potatoes and cocoa, just after the discovering of the Americas, or edible bird's nests and shark's fins in more recent times).

While certainly not exhaustive, this overview does reveal a real mythology of aphrodisiac foods, which oscillates between direct references to the symbolic—that is to say, "cultural"—dimension, on the one hand, and the exaltation of supposed "natural" effects deriving from foods, on the other hand. As the above-mentioned examples clearly show, these two levels are in fact often intertwined and in many cases not easily distinguishable from each other.

What is more, being food always inserted in particular practices and discourses, it is never simply a question of *natural* substances, but rather of culinary techniques and practices of preparation and consumption of those substances—that is to say, of cultural transformations of food matters and the meaning-making processes entailed by these. In Stano (2018), for instance, I focused on some relevant case studies referred to cinema to provide some insights on this crucial issue. An interesting case as related to the issues discussed here is the movie *Chocolat*, directed by Hallström (2000) and based on the homonymous novel by Harris (1999). As reported in a review by *The New York Times*,[41] the film revolves around the question: "can the liberating virtues of pleasure—here embodied by the warm, inviting power of chocolate—triumph over well-meaning but closed-minded small-town zealots?". Chocolate, in fact, is presented as the aphrodisiac food *par excellence*, which is capable of awakening dormant passions and corrupting the virtue of even the most upright and composed men. Two alternative systems are therefore confronted: a dominant one, based on a dogmatic and uncompromising religious morality, which is figurativised

[41] https://www.nytimes.com/2000/12/15/movies/film-review-candy-power-comes-to-town.html, 15 December 2020 (Accessed: 4 December 2022).

by the inflexibility of the Compte de Reynaud; and an alternative culture (a "non-culture" in the eyes of former), which is repeatedly presented as a supposed "nature" that cannot be tamed or subjected to any cultural canon. This is personified by both Vianne, whose endless wandering seems to be determined, not by coincidence, by a natural element (the icy North wind), and by chocolate itself, which is seemingly endowed with an intrinsic (i.e. "natural") aphrodisiac power. However, the movie insists precisely on the extreme variety of practices for producing and consuming chocolate, depicting Vianne's (non-)culture as the ability to perfectly master them. In fact, the woman is often represented in the act of crushing cocoa beans, adding milk and mixing the fluid thus obtained in large pots, then selecting the right shape for each compound and decorating every product as appropriate. What is more, she is able to understand which type of chocolate is suitable for each person, which allows her—and chocolate itself—to seduce them, that is to say, in a Greimassian perspective, to grasp their value-object and, thanks to this, to fit into their narrative programmes.[42]

A similar configuration can be identified in *Como agua para chocolate* (Arau 1992), whose title alludes to the idiomatic expression "like water for chocolate", which is generally used to describe a person in the throes of passion: "boiling", like the water needed to make hot chocolate.[43] The movie then insists precisely on cooking practices, which are presented as the means through which desire can move from the one cooking to food itself, and from the latter to the people eating it, in a sort or "contagion", or "aesthetic grasp" (cf. Landowski 2003), which penetrates the body, making it unavoidably succumb to this "alchemy" (i.e. a real magical art, and therefore evidently a cultural process, capable of "grasping" the "other" regardless of his/her intentions).

To conclude, therefore, there is no much point in insisting on the "natural" features and effects of food. These, in fact, are inextricably tied to the cultural practices determining their conformations and meanings, as well as the ways by which food presents itself to the eater's sight, as well as to their touch, smell, hearing and palate, in a synaesthetic, total, and necessarily all-encompassing experience. As Calefato (2006: 181) observed in relation to the pleasure connected to food lust, definitely, the more elaborate and luxurious is the manipulation (that is to say, the "culturalisation") of food, the greater can be the composite and erotic intensity of taste.

## 5.6 Nature on the Plate: Concluding Remarks

From a euphoric conception of cooking as a bridge between (the disorder of) nature and (the order of) culture, according to the famous definition introduced by Lévi-Strauss mentioned above, modernity has marked a shift first to a particular praise for

---

[42] For a detailed analysis, see Stano (2018): 51–62.

[43] In fact, in most countries in Central and Latin America, hot chocolate is prepared using water, not milk, as it was in ancient times.

fermentation (i.e. the "rotten", in Lévi-Straussian terms), that is, a more natural transformation of food, and more recently even to a real denial of any process of culturalisation of food matters (i.e. the "raw"). In fact, as the opposition between GMOs (and other technological innovations applied to food) and the organic paradigm examined above effectively shows, culture is no longer conceived as an "order", as it was in the systems studied by Lévi-Strauss, but rather as an "insub-ordination" (to the laws of nature, which it is thought to destabilise); it is no longer regarded as a form of science (in the Latin acceptation of *scientia*, that is to say, knowledge), but rather as a form of *incoscientia*, namely the dystopic realisation of a too pretentious humanity that does not recognise its own limits and dares to go beyond them, inevitably causing damage and death.

This was evident also in the case of the so-called "clean eating" movement, which highlighted a crucial contradiction: while trying to "uncook" food, by dysphoricising eating products and practices for the excesses attributed on their culturalisation, it ended up strengthening another culture of food, based on the decontextualization, oversimplification and exaggeration of its basic units—which seem to have replaced foods themselves as elements that are "good (or bad) to think with" (cf. Lévi-Strauss 1962; Contreras and Serra 2016).

If any aspect of culture can be admissible, it is only through a process of naturalisation, that is to say, by means of an operation of "crystallisation" that transforms it into a never-changing "tradition", closed to any form of change and innovation. This emerged clearly in the case of the Mediterranean diet, whose collective imaginary still tend to promote the idea of an immutable, static model to be protected from the "threats" of globalisation, rather than a heritage re-constructed day after day.

Finally, in other cases, such as the ones discussed in relation to the aphrodisiac foodsphere, culture seems to be basically neglected: a sort of alchemical potential is attributed on foods themselves, often disregarding the fact that they are always and necessarily part of specific practices of preparation and consumption which actively contribute to shape and continuously modify their meanings, and hence their possible effects on those who eat them. Moreover, as the history of aphrodisiac foods shows, symbolic associations can also play a crucial role in such dynamics.

In each and any case, in conclusion, what emerges is that real mythologies of "uncooking" do exist and invest food and nature with certain systems of values, social relations, religious beliefs, and ideological convictions. In this sense, therefore, "uncooking" too can be certainly considered—to paraphrase Lévi-Strauss (1965)—as a language in which societies (and especially the contemporary Western ones) unconsciously translate their structure.

## References


Adamiec, Camille. 2013. Alimentation santé: de la prescription alimentaire à l'éthique de vie. La construction progressive des représentations du manger sain. *E|C* VII (14): 59–68.

Angela, Alberto. 2012. *Amore e sesso nell'antica Roma*. Milan: Mondadori.

Bertero, Alessandro, Stefano Biressi, Francesco Buscemi, Luciano Conti, Matteo Cresti, Cesare Gargioli, Luca Lo Sapio, Barbara Lucia Loera, Cristina Poncibò, and Simona Stano. 2023. We need an informed discussion on cultivated meat. Technical, social, and humanities specialists call for measured discussion on the future of food. *Nature Italy*, Apr 17. https://www.nature.com/articles/d43978-023-00056-1. Accessed 30 June 2023.

Boutaud, Jean-Jacques. 2013. Dimagrire in regime mediatico. In *Dietetica e semiotica. Regimi di senso*, ed. Gianfranco Marrone and Dario Mangano, 19–37. Milan-Udine: Mimesis.

Bratman, Steven. 1997. The health food eating disorder. *Yoga Journal* (September-October), 42–50.

Bratman, Steven, and David Knight. 2001. *Health Food Junkies. Orthorexia Nervosa: Overcoming the Obsession with Healthful Eating*. New York, NY: Broadway.

Bressanini, Dario, and Beatrice Mautino. 2015. *Contro natura: Dagli OGM al "bio", falsi allarmi e verità nascoste del cibo che portiamo in tavola*. Milan: Rizzoli.

Buscemi, Francesco. 2015. New meat and the media conundrum with nature and culture. *Lexia* 19–20: 419–434.

Buosi, Chiara. 2004. La cucina d'autore. Un'analisi sociosemiotica delle cucine di La Madonnina del pescatore, L'Esguard, El Bulli. *E|C*, 1–24. http://www.ec-aiss.it/includes/tng/pub/tNG_download4.php?KT_download1=339e7338596a6ba4b9249029dba8f319. Accessed 30 Nov 2022.

Calefato, Patrizia. 2006. Lussi e lussurie alimentari. In *Semiofood. Comunicazione e cultura del cibo*, ed. Giovanni Manetti, Paolo Bertetti and Alessandro Prato, 175–191. Torino: Centro Scientifico Editore.

Camporesi, Piero. 1989. *I balsami di Venere. L'erotismo in Europa dal Medioevo al Settecento*. Milan: Garzanti.

Contreras Hernández, Jesús, and Joan Ribas Serra. 2016. Are nutrients also good to think? *Semiotica* 211: 139–163.

Corrett, Natasha and Vicki Edgson. 2012. *Honestly Healthy: Eat with Your Body in Mind, the Alkaline Way*. London, UK: Jacqui Small LLP.

Danieli, Francesco. 2019. "Per eccitar Venere" o "Per dirizzar la verga". La dieta afrodisiaca nei trattati leonardiani sul cibo dell'amore. *Idomeneo* 28: 157–164.

Davis, William. 2011. *Wheat Belly: Lose the Wheat, Lose the Weight, and Find Your Path Back to Health*. New York, NY: Rodale.

Davis, William. 2014. *Wheat Belly Total Health: The Ultimate Grain-Free Health and Weight-Loss Life Plan*. New York, NY: Rodale.

Del Vicario, Michela, Gianna Vivaldo, Alessandro Bessi, Fabiana Zollo, Antonio Scala, Guido Caldarelli, and Walter Quattrociocchi. 2016. Echo chambers: Emotional contagion and group polarization on facebook. *Scientific Reports* 6: 37825. https://www.nature.com/articles/srep37825. Accessed 30 Nov 2022.

Douglas, Mary. 1972. Deciphering a meal. *Daedalus* 101 (1): 61–81. Reprinted in Id. 1975. *Implicit Meanings*, 249–275. London: Routledge

EUFIC (European Food Information Council). 2012. *What are superfoods and are they really super?* https://www.eufic.org/en/healthy-living/article/the-science-behind-superfoods-are-they-really-super. Accessed 30 Nov 2022.

FAO (Food and Agricultural Organization). 1999. Organic agriculture. *Report of the Committee on Agriculture*, Fifteenth Session. Rome, Jan 25–29. https://www.google.com/url?sa=t&rct=j&q=&esrc=s&source=web&cd=&ved=2ahUKEwjrrqu4wtj7AhX-iv0HHYZQDPwQFnoECAwQAQ&url=https%3A%2F%2Fwww.fao.org%2Funfao%2Fbodies%2Fcoag%2Fcoag15%2Fdocs%2FX0075E.doc&usg=AOvVaw3a665Q4g_lUPVWp0pJbRMC. Accessed 30 Nov 2022.

Ferraro, Guido. 1998. *L'emporio dei segni*. Rome: Meltemi.

Fivian, Emily, and Carly J. Wood. 2019. The roles of social media, clean eating and self-esteem in the risk of disordered eating: A pilot study of self-reported healthy eaters. *International Journal of Food, Nutrition and Public Health* 10: 28–39.
Harris, Joanne. 1999. *Chocolat*. London: Doubleday.
Greimas, Algirdas J. 1966. *Sémantique structurale*. Paris: Larousse.
Greimas, Algirdas J. 1983. La soupe au pistou ou la construction d'un objet de valeur. In *Du sens II*, 157–169. Paris: Seuil.
Habermas, Jürgen. 1971 [1962]. *Strukturwandel der Öffentlichkeit. Untersuchungen zu einer Kategorie der bürgerlichen Gesellschaft. 5*. Neuwied/Berlin: Auflage. English Trans. Burger, Thomas, with the assistance of Lawrence, Frederick. 1991. *The Structural Transformation of the Public Sphere: An Inquiry into a Category of Bourgeois Society*. Cambridge, MA: The MIT Press.
Hobsbawm, Eric, and Terence Ranger. 1983. *The Invention of Tradition*. Cambridge, MA: Cambridge University Press.
Ieranò, Giorgio. 2019. *Il mare d'amore. Eros, tempeste e naufragi nella Grecia antica*. Bari-Rome: Laterza.
Junger, Alejandro. 2009. *Clean: The Revolutionary Program to Restore the Body's Natural Ability to Heal Itself*. New York, NY: HarperOne.
Keys, Ancel. 1980. *Seven Countries. A Multivariate Analysis of Death and Coronary Heart Disease*. Cambridge, MA: Harvard University Press.
Landowski, Eric. 2003. *Passions sans nom. Essais de socio-sémiotique III*. Paris: PUF.
Laudan, Rachel. 2004 [2000]. Birth of the modern diet. *Scientific American* 283 (2): 62–67.
Lawson, Nigella. 2015. *Simply Nigella: Feel Good Food*. Chicago, IL: Flatiron Books.
Lévi-Strauss, Claude. 1962. *Le totémisme aujourd'hui*. Paris: PUF. English Trans. Needham, Rodney. 1963. *Totemism*. Boston, MA: Beacon Press.
Lévi-Strauss, Claude. 1964. *Mythologiques I. Le cru et le cuit*. Paris: Plon. English Trans. Weightman, John and Doreen. 1969. *The Raw and the Cooked*. New York, NY, and Evanston, IL: Harper & Row, Publishers.
Lévi-Strauss, Claude. 1965. Le triangle culinaire. *L'Arc* 26: 19–29. Reprinted in *Le Nouvel Observateur Hors-Série* (2009), 14–17. http://palimpsestes.fr/textes_philo/levi_strauss/triangle_culinaire.pdf. Accessed 1 Dec 2022. English Trans. Brooks, Peter. 2013. The culinary triangle. In *Food and Culture: A Reader*, ed. Carole Counihan and Penny Van Esterik, 40–47. New York, NY, and London, UK: Routledge.
Lévi-Strauss, Claude. 1968. *Mythologiques III. L'origine des manières de table*. Paris: Plon. English Trans. Weightman, John and Doreen. 1978. *The Origin of Table Manners*. New York, Hagerstown, San Francisco, CA: Harper & Row.
Lotman, Jurij M. 1984. О семиосфере. *Труды по знаковым системам (Signs Systems Studies)* 17: 5–23. English Trans. Clark, Wilma. 2005. On the semiosphere. *Signs Systems Studies* 33 (1): 205–229.
Lyman, Eric. 2000. Pope expresses opposition to GMOs: Cites need for 'the respect of nature'. *Bureau of National Affairs* 221. http://online.sfsu.edu/~rone/GEessays/PopeGMO.htm. Accessed 14 Nov 2020.
Marrone, Gianfranco. 2011. *Addio alla natura*. Torino: Einaudi.
Marrone, Gianfranco. 2017. Cibo e linguaggio. Traduzioni e tradizioni. In *Parole per mangiare. Discorsi e culture del cibo*, ed. Irina Bajini, Maria Vittoria Calvi, Giuliana Garzone and Giuseppe Sergio, 19–34. Milan: LED.
Martinis, Bruno. 1999. *I fiori degli dei. Le erbe in magia e medicina*. Bari: Dedalo.
McCartney, Margaret. 2016. Clean eating and the cult of healthism. *British Medical Journal* 354: i4095.
Montanari, Massimo. 2016. *Mangiare da cristiani: diete, digiuni, banchetti. Storie di una cultura*. Milan: Rizzoli.
Nicolosi, Guido. 2006–2007. Biotechnologies, alimentary fears and the orthorexic society. *Tailoring Biotechnologies* 2 (3): 37–56.

Nicolosi, Guido. 2007. *Lost food. Comunicazione e cibo nella società ortoressica*. Catania: Ed.it.
Niola, Marino. 2012. *Miti d'oggi*. Milan: Bompiani.
Niola, Marino. 2014. Prefazione. In Moro, Elisabetta, *La dieta mediterranea. Mito e storia di uno stile di vita*, 7–11. Bologna: Il Mulino.
Niola, Marino. 2015. *Homo dieteticus. Viaggio nelle tribù alimentari*. Bologna: Il Mulino.
Ortoleva, Peppino. 2019. *Miti a bassa intensità. Racconti, media, vita quotidiana*. Torino: Einaudi.
Perlmutter, David. 2013. *Grain Brain: The Surprising Truth About Wheat, Carbs, and Sugar—Your brain's silent killers*. New York, NY: Little, Brown and Company.
Pollan, Michael. 2008. *In Defense of Food: An Eater's Manifesto*. New York, NY: Penguin Press.
Quattrociocchi, Walter, and Antonella Vicini. 2016. *Misinformation. Guida alla società dell'informazione e della credulità*. Milan: Franco Angeli.
Reno, Tosca. 2007. *The Eat-Clean Diet*. Mississauga, ON: Robert Kennedy Publishing.
Rodale, Maria. 2010. *Organic Manifesto: How Organic Food Can Heal Our Planet, Feed the World, and Keep Us Safe*. Emmaus, PA: Rodale Press.
Scrinis, Gyorgy. 2008. On the ideology of nutritionism. *Gastronomica* 8 (1): 39–48.
Scrinis, Gyorgy. 2013. *Nutritionism: The Science and Politics of Dietary Advice*. New York, NY: Columbia University Press.
Shelley, Mary. 1818. *Frankenstein, or, The Modern Prometheus*. London: Lackington, Hughes, Harding, Mavor, & Jones.
Stano, Simona. 2015a. *Eating the Other. Translations of the Culinary Code*. Newcastle-upon-Tyne: Cambridge Scholars Publishing.
Stano, Simona. 2015b. From nutrients to foods: The alimentary imaginary of the mediterranean diet. *Essachess—Journal for Communication Studies* 8, 2 (16): 115–132.
Stano, Simona. 2016. Tell me what you *do not* eat, and i shall tell you what you are. Food, health, and conspiracy theories. *Lexia* 23–24: 327–343.
Stano, Simona. 2018. *I sensi del cibo. Elementi di semiotica dell'alimentazione*. Rome: Aracne.
Stano, Simona. 2021a. Beyond nutrition: Meanings, narratives, myths. In *Food for Thought. Nourishment, Culture, Meaning*, ed. Simona Stano and Amy Bentley, 147–158. Cham: Springer.
Stano, Simona. 2021b. Food, health and the body: A biosemiotic approach to contemporary eating habits. In *Food and Medicine: A Biosemiotic Perspective*, ed. Victoria N. Alexander, Yogi H. Hendlin and Jonathan Hope, 43–69. Cham: Springer.
Stano, Simona. 2021c. Il discorso alimentare tra teorie cospirative e mitologie del presente digitale: un approccio semio-culturale. In *Trame nascoste. Teorie della cospirazione e miti sul lato in ombra della società*, ed. Nicola Pannofino and Davide Pellegrino, 49–67. Milan-Udine: Mimesis.
Szejda, Keri, Courtney Dillard, and Tessa Urbanovich. 2019. Initial consumer perceptions of cellular agriculture nomenclature: A qualitative analysis of word associations. *GFI Academic papers*. https://gfi.org/images/uploads/2020/01/Dec-2019-CM-Nomenclature-Qual-Study-Word-Associations-FINAL-REPORT.pdf?pi_content=6be253415f92d884a69c564ba671b6d2dd3224232c29250fbf5fe78da0898dd0. Accessed 14 June 2023.
Tandoh, Ruby. 2016. The unhealthy truth behind 'wellness' and 'clean eating'. *VICE*, May 13. https://www.vice.com/en/article/jm5nvp/ruby-tandoh-eat-clean-wellness. Accessed 30 Nov 2022.
Tylor, Edward B. 1871. *Primitive Culture: Researches into the Development of Mythology, Philosophy, Religion, Art, and Custom*. London: John Murray.
UNESCO. 2012. *Mediterranean diet*. http://www.unesco.org/culture/ich/en/RL/00884. Accessed 14 Nov 2022.
UNESCO. 2013. *Mediterranean diet*. http://www.unesco.org/culture/ich/RL/00884. Accessed 14 Nov 2020.
Valnet, Jean. 2010 [1985]. *Se soigner par les légumes, les fruits et les céréales*. Paris: Ed. Libraire Maloine.
Ventura Bordenca, Ilaria. 2012. Nature in vendita. Il packaging dei prodotti biologici. In *Semiotica della natura (Natura della semiotica)*, ed. Gianfranco Marrone, 277–306. Milan-Udine: Mimesis.

Ventura Bordenca, Ilaria. 2015. "Questa non è una dieta". Linguaggi e strategie del dimagrimento. In *Buono da pensare*, ed. Gianfranco Marrone, 199–236. Rome: Carocci.
WHO (World Health Organization). 2014. *3. Is the safety of GM foods assessed differently from conventional foods? Frequently asked questions on genetically modified foods*. http://www.who.int/foodsafety/areas_work/food-technology/faq-genetically-modified-food/en/. Accessed 12 Nov 2022.
Wilson, Bee. 2017. Why we fell for clean eating. *The Guardian*, Aug 11. https://www.theguardian.com/lifeandstyle/2017/aug/11/why-we-fell-for-clean-eating. Accessed 29 Nov 2022.
Woodward (Mills), Ella. 2015. *Deliciously Ella*. London: Yellow Kite—Hachette UK.
Woodward (Mills), Ella. 2016. *Deliciously Ella Every Day*. London: Yellow Kite—Hachette UK.
Woodward (Mills), Ella. 2017. *Deliciously Ella With Friends*. London: Yellow Kite—Hachette UK.
Wynne, Lauren A., Gareen Hamalian, and Neve Durrwachter. 2021. Virtue and disease: narrative accounts of orthorexia nervosa. In *Food for Thought. Nourishment, Culture, Meaning*, ed. Simona Stano and Amy Bentley, 171–182. Cham: Springer.

## Filmography

Arau, Alfonso. 1992. *Como agua para chocolate*. Mexico, 105'.
EXPO 2015. 2015. *La dieta mediterranea tornerà mai a casa?* Italy, 1' 49". https://www.youtube.com/watch?v=l9_lDREniYU. Accessed 30 Nov 2022.
Hallström, Lasse. 2000. *Chocolat*. United Kingdom and United States, 121'.
Paparusso, Nicola. 2013. *Dieta mediterranea—Ambasciatrice della salute nel mondo*. Italy, 2' 23". https://www.youtube.com/watch?v=NRI3Q3ESKN8. Accessed 30 Nov 2022.
Quinn, Tristan. 2017. *Clean Eating: The Dirty Truth—Horizon (BBC)*. United Kingdom, 58' 21". https://archive.org/details/BBCHorizonCleanEatingTheDirtyTruthFullDocumentary. Accessed 30 Nov 2022.

# Chapter 6
# "Just as Nature Intended": Reflections on Nakedness and Corporeality

**Abstract** At first glance, the naked body seems to recall the alleged materiality of the "Me-flesh", conceived as an extensive reality of the "natural" world: we are all born naked, "just as nature intended"—as the popular expression goes. However, since the very first moment it shows itself to the outside world, our body is "dressed up" with signs, that is to say, it is marked by specific rituals and systems of values that are inherent to the semiosphere it becomes part of. In fact, nudity has been variously signified in different periods and cultures, as this chapter shows. Moreover, the naked body has been largely used as a powerful tool in political and social action, artistic performance, and a series of other significant practices aimed at (re)writing and (re)semantising the corporeal dimension for different purposes. The analysis of relevant case studies explores such dynamics, providing insights on the ambivalence of the naked body between its supposed natural ascription and its multiple cultural characterisations.

## 6.1 Nakedness in Between Nature and Culture

At first glance, the naked body seems to recall the alleged materiality of the "Me-flesh" (i.e. le *corps-chair* described by Fontanille 2004), conceived—in Cartesian terms (Descartes 1637)—as an extensive reality of the "natural" world: we are all born naked, "just *as nature* intended"—as the popular expression goes.

In this respect, it is particularly interesting to consider the myth of Epimetheus described in Plato's *Protagoras* (320d–322a):

> Once upon a time there were gods only, and no mortal creatures. But when the time came that these also should be created, the gods fashioned them out of earth and fire and various mixtures of both elements in the interior of the earth; and when they were about to bring them into the light of day, they ordered Prometheus and Epimetheus to equip them, and to distribute to them severally their proper qualities. Epimetheus said to Prometheus: 'Let me distribute, and do you inspect.' This was agreed, and Epimetheus made the distribution. There were some to whom he gave strength without swiftness, while he equipped the weaker with swiftness; some he armed, and others he left unarmed; and devised for the latter some other means of preservation, making some large, and having their size as a protection, and others small, whose nature was to fly in the air or burrow in the ground; this was to be their way of escape. Thus did he compensate them with the view of preventing any race from



S. Stano, *Critique of Pure Nature*, Numanities - Arts and Humanities in Progress 26,
https://doi.org/10.1007/978-3-031-45075-4_6

> becoming extinct. And when he had provided against their destruction by one another, he contrived also a means of protecting them against the seasons of heaven; clothing them with close hair and thick skins sufficient to defend them against the winter cold and able to resist the summer heat, so that they might have a natural bed of their own when they wanted to rest. ... Thus did Epimetheus, who, not being very wise, forgot that he had distributed among the brute animals all the qualities which he had to give,—and when he came to man, who was still unprovided, he was terribly perplexed. Now while he was in this perplexity, Prometheus came to inspect the distribution, and he found that the other animals were suitably furnished, but that man alone was naked and shoeless, and had neither bed nor arms of defence (Engl. Trans. in Williams 2009: 121).

While every animal is covered by fur, feathers and other "positive traits" protecting it (from other animals, as well as from other possible threats in the surrounding environment), man remains "naked", and therefore extremely vulnerable (cf. Ariemma 2006, 2007).

However, as Jacques Derrida remarked, it is not the mere fact of having his body uncovered that actually makes man naked, but rather his *consciousness* of this:

> It is generally thought ... that the property unique to animals, what in the last instance distinguishes them from man, is their being naked without knowing it. Not being naked therefore, not having knowledge of their nudity, in short, without consciousness of good and evil. From that point on, naked without knowing it, animals would not be, in truth, naked. They wouldn't be naked because they are naked. In principle, *with the exception of man, no animal has ever thought to dress itself* (2006: 39: Engl. Trans. 2008: 4–5, my emphasis).

In fact, since the very first moment it shows itself to the outside world, our body is "dressed up" with signs, that is to say, marked by specific rituals and systems of values (e.g. clothing, medical practices, aesthetic standards, regulatory codes, etc.) that are inherent to the cultural context it is part of and to which it allows us to connect. This is precisely what makes it emerge as both an object in the world and a way of access to that world; a natural substance and at the same time an organising—and organised,—form; an interpreted object and an interpreting subject; in other terms, not only something which is signified, but also a signifying entity actively participating in the processes of meaning-making—of the world, of other bodies, and of itself. Not by chance, nudity has also been largely used as a powerful tool in political action (see in particular Brown and Gershon 2020; Brownie 2017; Lunceford 2012, 2019), artistic performance (see in particular Chare and Contogouris 2022; Lu 2007; Westcott 2010; Abramović et al. 2010), and a series of other significant practices aimed at (re)writing and (re)semantising the corporeal dimension for different purposes.

In order to explore such an ambivalence of the naked body between its supposed natural ascription and its multiple cultural characterisations, the following paragraphs deal with an analysis of nakedness across different ages and contexts, paying particular attention to specific collective imaginaries and processes of (re)semantisation of nakedness.

## 6.2 Nudity in the Classical Age

Nudity was rather common for men in Ancient Greece, especially in the athletic sphere. Initially wearing a loincloth, Greek athletes began to compete naked at a particular moment in time,[1] as Plato supports in the *Republic* (452C), expressing implicit praise for such a change:

> Not long ago, as we shall remind them, the Hellenes were of the opinion, which is still generally received among the barbarians, that the sight of a naked man was ridiculous and improper; and when first the Cretans and then the Lacedaemonians introduced the custom, the wits of that day might equally have ridiculed the innovation. … But when experience showed that to let all things be uncovered was far better than to cover them up, and the ludicrous effect to the outward eye vanished before the better principle which reason asserted, then the man was perceived to be a fool who directs the shafts of his ridicule at any other sight but that of folly and vice, or seriously inclines to weigh the beautiful by any other standard but that of the good (Engl. Trans. 1892).

Other authors, such as Herodotus,[2] also insisted on nakedness as a distinctive trait of the Greeks (especially as opposed to barbarians, whose refusal to get naked they saw as a shame, and a sign of tyranny and political repression), marking its essential role in asserting freedom, privilege and physical virtues (see Kyle 2015). Moreover, athletes "were thought to be protected in some way by their nudity" (Bonfante 1975: 102), as this was believed to inspire fear and horror in their adversaries (Mouratidis 1985: 221), as well as to have apotropaic effects and other magical properties (see Burkert 1979).

On the contrary, female nudity appeared rarely, and under very specific circumstances. As Bonfante (1989: 558–562) noted, in archaic arts the female nude appeared only occasionally, as a religious fertility motif, probably following the Near Eastern model of the naked Mother Nature goddess. Afterwards, it was mainly used "for thetic appeal, for magic or erotic appeal" (*Ibid.*: 558), and more generally for representing prostitutes (especially in Attic vase painting):

> Respectable women did not go out much, they did not attend male symposia, and they certainly did not undress in public. They were in fact protected from the sun, from men's eyes, and from the evil eye by dresses and mantles that covered them from head to foot (*Ibid.*: 559).

With just a few exceptions, in sum, in Greek literature and figurative representations, as in everyday life, a widespread taboo against female nakedness in public

---

[1] Drawing on an epitaph on Orsippus ("First of the Greeks in Olympia was he crowned while naked; Before him, all contestants were girdled in the stadium") attributed to Simonides (see Hicks 1882: 1), Arieti supports that it was "during the 15th Olympic games in 720 B.C., when the loincloth flew from Orsippus' waist, the Greeks cast aside their modesty. It is not known whether the cloth fell by chance, nor whether Orsippus with pure or mischievous intent threw it aside, but he won two first prizes that day; he was victor in the race, and the first to be crowned naked. … Never again would a male Greek athlete take up the cloth" (Arieti 1975: 431).

[2] "For among the Lydians, as among the barbarians generally, to be seen naked, even for a man, is considered a great shame" (1.10.9–11), quoted in Bonfante (1989: 546).

remained in full force. Among the exceptions, the case of goddesses, and especially of Aphrodite of Knidios, deserves particular attention. In fact, the first Greek complete nude female in sculpture is believed to be a statue of the goddess (which, it is worthwhile reminding, was the deity of sexual love and beauty) sculpted by Praxiteles—with some authors suggesting an influence of earlier representations of Aphrodite's precursor, the Mesopotamian goddess Astarte (see Bahrani 1996), while others pointing to the role played by classical male nudes in its creation (see Salomon 1997). However, as Nanette Salomon (*Ibidem*) noted, unlike nude male statues (consider, for instance, the popular sculpture of Hermes and Dionysus by Praxiteles himself), Aphrodite's right hand generally covered her pubic region, recalling a *pudica* gesture of shame and discomfort—which is symptomatic of the mentioned taboo.[3]

Moving from Greek traditions to the later Etruscan and Roman athletics, public nudity was banned for men too, as it became associated with homoeroticism.[4] In the Roman empire, the *toga* became a fundamental element of both daily experience and artistic representations, signalling the status and rank of citizens (see Habinek and Schiesaro 1997). On the contrary, as reported by Ennius, "exposing naked bodies among citizens [wa]s the beginning of public disgrace",[5] a sentiment echoed by Cicero, who also explicitly linked self-containment and modesty with citizenship. Connected with imprisonment (since captives were stripped) and slavery (since slaves for sale were often naked), nudity therefore took on a marked negative connotation, for both men and women. However, the influence of Greek art led to some exceptions, such as (mainly partial) nude portraits of Roman gods and men, or symbolic references (e.g. the choice of showing off breast as a reference to the ideas of nurturing and abundance, as described in Chap. 3).

## 6.3 Nudity in Christian Culture

Another crucial point in the history of nudity is the advent of Christianism. While pagan nudity was associated with sex, Christian nudity was usually related to grace (Ariès and Duby 1987), as well exemplified by the act of baptism, which originally consisted in the full immersion of the naked body in dedicated basins attached to the cathedrals.

Even in the famous passage of *Genesis* concerning Adam and Eve's sin and its consequences, nudity is not stigmatised as such. In fact, its negative connotation is

[3] In this respect, it is also important to mention that no direct reference to female genitals can be found in literature either, as they are generally called *aidoia*, literally meaning "shameful parts" (cf. Bahrani 1996: 6), with reference to Aidos or Aedos, the Greek goddess of shame, modesty, respect and humility.

[4] A possible exception was represented by the baths, where public nudity was accepted. However, various sources in literature show that attitudes towards nude bathing also changed over time.

[5] "Flagitii principium est nudare inter civis corpora", quoted by Cicero (*Tusc. Disp.* 4.70).

explicitly related to the experience of the forbidden fruit, which makes Adam and Eve become aware of being naked and hence ashamed of their condition:

> Then the eyes of both of them were opened, and they realized they were naked; so they sewed fig leaves together and made coverings for themselves. Then the man and his wife heard the sound of the Lord God as he was walking in the garden in the cool of the day, and they hid from the Lord God among the trees of the garden. But the Lord God called to the man, "Where are you?" He answered, "I heard you in the garden, and I was afraid because I was naked; so I hid." And he said, "Who told you that you were naked? Have you eaten from the tree that I commanded you not to eat from?" (*Genesis* 3: 7–11).

While acquiring awareness ("the eyes of both of them were opened"), Adam and Eve were subjected to a process of "deprivation": as interestingly remarked by Giorgio Agamben (in turn recalling Augustine of Hippo's *De civitate Dei*), originally, in fact, they were not in a condition of nudity, but covered by a "garment of grace" (see Agamben 2009: 86). Becoming aware of their nudity, they lost such a garment and became exposed and vulnerable, thus feeling the need to cover their body.[6] This aspect was particularly emphasised in Medieval art, up until to 15th century paintings (cf. Sorabella 2008)—as illustrated, for instance, by the scene represented in *The Creation and the Expulsion from the Paradise* (c. 1445) by Giovanni di Paolo.

The process of dysphorisation of nudity was strengthened with the discovery of the so-called "New World", as attested by various conquerors and explorers' reports and works (e.g. Gonzalo Fernández de Oviedo y Valdès' *Sumario de la natural historia de las Indias*, 1526), which described the inhabitants of such unexplored lands as "savages", and their nudity as a sign of "immorality" and "animality". A change was then registered towards the 16th and the 17th centuries, when nakedness became the symbol of the "good savages", representing their "state of nature"—as supported, for instance, in Jean-Baptiste DuTertre's *Histoire générale des îles Saint-Christophe, de la Guadeloupe, de la Martinique et autres de l'Amérique* (1654), which insisted precisely on the fact that the natives' sole garment was "the one provided to them *by nature*" (my emphasis), thus marking the opposition with European "culture" and habits.

## 6.4 Nudity from the Renaissance to Contemporary Culture

The rediscovery of classical culture in the Renaissance restored the nude in figurative arts:

> Nude figures based on antique models appear in Italy as early as the mid-thirteenth century, and by the mid-fifteenth century, nudes had become symbols of antiquity and its reincarnation. Donatello adapted the idealized proportions of Greek athletic figures for his celebrated

[6] In this respect, Augustine wrote: "But when they were stripped of this grace, that their disobedience might be punished by fit retribution, there began in the movement of their bodily members a shameless novelty, which made nakedness indecent: it at once made them observant and made them ashamed" (*De civitate Dei*, 14.17).

> statue of David (ca. 1440; Bargello, Florence) and thus presented a biblical hero in classical guise. In a widely circulated engraving, Pollaiuolo used nude figures in vigorous poses to suggest the range of human action [ca 1470–90]. In the next generation, Michelangelo made his own colossal statue of David, again conceived as an antique nude (1501–4; Accademia, Florence), and elsewhere he devoted unique artistic energy to the male nude. His enthusiasm for the subject was such that he introduced nudes even in religious paintings, including the frescoes of the Sistine Chapel, and he used studies of the male form to imbue figures of every sort with Herculean massiveness and power (Sorabella 2008).

Female nudes also appeared, as exemplified by various representations of Venus by Titian (e.g. *Venus and the Lute Player*, ca 1565–70, *Venus of Urbino*, 1538, *Venus and Adonis*, 1550s) and other Venetian painters, which depict her while lying naked in a landscape or domestic interior. Moreover, a nude type for the Christ Child was developed, with genitals exposed to express the theological status of Christ as God made man (see Steinberg 1983).

Nonetheless, the taboo of nudity remained in the background both in the arts[7] and in everyday life, evidently growing again towards the 19th century, when strict disciplinary codes (such as the adoption of swimwear covering up to the knee, and the impossibility, even for men, to show themselves bare-chested on the beach) were applied to the body, remaining in force until the following century. It must be noted, however, that artistic nudes remained a constant presence, as evidenced by a number of famous works by Romanticists such as Jean-Auguste Dominique Ingres (e.g. *Baigneuse de Valpinçon*, 1808, *La grande odalisque*, 1814, *Le bain turc*, 1862, and *Vénus anadyomène*, 1848), Théodore Géricault (*Le Radeau de la Méduse*, 1818–19, *Léda et le Cygne*, 1822) and Eugène Delacroix (*La Barque de Dante*, 1822, *Mademoiselle Rose (Seated Nude)*, 1824, *La femme aux bas blancs*, 1825–30, *Le Lever*, 1850), Realists like Gustave Courbet (e.g. *Les Baigneuses*, 1853, *La Femme au perroquet*, 1865, *La source*, 1868, and especially *L'Origine du monde*, 1866, which caused much scandal), Jean-Baptiste Camille Corot (*Nymphe couchée à la campagne*, 1857–59, *Bacchante couchée au bord de la mer*, 1865, *Marietta (L'Odalisque romaine)*, 1843) and John Singer Sargent (*A Nude Boy on a Beach*, 1878), and Impressionists such as Edouard Manet (e.g. *Déjeuner sur l'herbe*, 1863) and Pierre-Auguste Renoir (*Grand nu*, 1907, *Nu couché, vu de dos*, 1916).

With the spread of cultural relativism in the last decade of the 19th century and the first decades of the 20th century, different meanings and values of nudity were finally recognised, in light of the specific cultural contexts within which it was adopted. From a sign of "primitiveness", and sometimes even "immorality", the naked body was thus fully admitted as a polysemantic cultural trait.[8] Accordingly, representations of the Other's naked body became more and more frequent in the arts, as well reflected

---

[7] In this respect, Titian's paining *Amor sacro e amor profano* (1514) is particularly interesting, as it uses nudity precisely to mark the distinction between "sacred love" (symbolised by the woman on the left, fully and richly dressed) and "profane love" (represented by the woman on the right, completely naked except for a white cloth covering her loins and a large red mantle put over her left shoulder).

[8] Hence nudity emerged as a means for signalling power and social distinctions. In *Tangba* villages, in Benin, as it is common in West Africa, for instance, nakedness is a way of affirming one's identity as indigenous, as opposed to people of foreign origin. In fact, clothes themselves are of foreign

by Paul Gaugin's works (e.g. *Manao tupapau (L'Esprit des morts veille)*, 1892, *Aha Oe Feii (Eh quoi ! Tu es jalouse ?)*, 1892, *Parau na te Varua ino (Paroles du diable)*, 1892, *Vairumati tei Oa (Femme tahitienne assise et idole)*, 1892, *Eu haere ia oe (Où vas-tu ?)*, 1893, *Hina te Fatou (La lune et la Terre)*, 1893, and *Hannah la Javanaise*, 1894).

In the meantime, social nudity began being celebrated as "a way of life in harmony with nature … with the intention of encouraging self-respect, respect for others and for the environment" (INF-FNI 1974 in Deschênes 2016: 2), with a marked connection between nudism and the desire to "reintegrate man into nature"—as remarked also by the name chosen for such a lifestyle: *naturism*. Major promoters of this idea included Heinrich Pudor (who in 1906 published a three-volume treatise entitled *Katechismus der Nackt-Kultur*, where he presented the benefits of nudity in co-education and advocated participating in sports while "being free of cumbersome clothing", thus recalling Greek habits), Richard Ungewitter (who in his works *Die Nacktheit* (1906) and *Nackt* (1921 [1908]) claimed the importance of combining the nudist philosophy with fitness and fresh-air bathing for good physical and mental health, as well as for an improved moral-life view) and several others thinkers and physicians (such as Gaston and André Durville, who stressed the importance of "naturisme" as a lifestyle based on the consumption of "natural foods" and a "natural" relation with the environment, including nudity in the latter). Organised clubs for nudists were opened first in Germany (consider, for instance, *Freilichtpark*, the Free-Light Park inaugurated near Hamburg in 1903 by Paul Zimmerman) and various national federations united to form the International Naturist Federation (INF-FNI), making naturism spread throughout Europe and to the United States, where it became established in the 1930s, and continued growing across the century.

The phenomenon then progressively lost its importance in the last decades of the 20th century. However, it initiated a general process of liberalisation of nudity both in everyday life and media discourses,[9] with an unprecedented process of exposition of the body which, in the last decades, has brought about new aesthetic codes and possibilities of empowerment, therefore emerging as a mighty "political" means.[10] I will deal with the most relevant aspects of such a process in the following paragraphs.

---

origin, having been imported first from Islamic countries and later from Europe, while nudity is seen as a habit—i.e. a cultural trait—of the ancestors.

[9] Not without exceptions, as exemplified by the limitations imposed by the Nazi regime, which banned most independent naturist organisations, as well as by Futurists' aversion to nudes.

[10] In fact, the concealment of the body has rather been increasingly condemned as a form of "primitiveness" and "backwardness", as current controversies such as the French debate on the so-called "burkini" show (cf. Almeida 2018).

## 6.5 Nudity as a "Political" Means, Between Protest and Performance Art

From "flower children" adopting a naturalist lifestyle to challenge sociocultural norms to feminist and environmental activists, from participants in the World Naked Bike Ride demonstrating against car culture to young female college students deliberately exhibiting their unveiled bodies on Collegehumor.com (see Lunceford 2012), nudity has largely proved to be a powerful means for social change and political action. While dating back as far as to the times of Diogenes, who showed himself naked in public as an affront to commonly accepted behaviours (*Ibid.*: 2), this phenomenon has become more evident in the 20th and especially the 21st century, mainly as related to political protest and artistic performance.

The following paragraphs will focus on some particularly relevant case studies, in order to highlight the main uses and valorisations of nakedness, as well as the effects of meaning deriving from them, especially as related to the ambivalence of the corporeal dimension between its supposed "nature" and its "cultural" characterisation.

### 6.5.1 *The Naked Athena: Nudity, Protest, and Empowerment*

Wearing only a hat and a black mask to partially cover her face, the "Naked Athena"—as she has been "baptised" by the media—sits on the ground with her legs apart, motionless in front of a number of police officers in downtown Portland, Oregon. Among the images of the protests linked to the *Black Lives Matter* movement, this one, which was taken in July 18 2020 by Dave Killen,[11] and later shared by thousands of users on Twitter and other social media, is undoubtedly one of the most evocative.

A series of significant factors, including a particular role attributed on the naked body, characterise it. In fact, the differential character of nudity is central in this case, as remarked by the perspective chosen by the photographer, which emphasises the contrast between the woman's naked body and the other, almost completely covered, bodies surrounding her. Such a contrast entails a series of crucial semantic oppositions:

- "openness" (of the female figure, who purposedly sits—and then, when she is forced to move, also stands—with her arms and legs open to ensure a full "exposure" of her naked body) *versus* "closeness" (of the police officers, restrained and gathered to form an encircling barricade);
- "uniqueness" (the Naked Athena distinguishes herself from everyone else around precisely through the exposition of her body) *versus* "seriality" (as highlighted by the chromatic and eidetic regularity of the police officers' uniforms, which makes them practically indistinguishable from each other);

[11] For the picture and a short report on the events, see in particular Nguyen (2020).

- "singularity" (as related to the protester's uniqueness) *versus* "indistinct multiplicity" (as seriality contributes to highlight the compactness of the barricade, making it prevail over the individuality of the people forming it);
- "femininity" (nudity clearly reveals the gender identity of the exposed body) *versus* "a-sexuality" (by contrast, the uniforms conceal that of the police officers);
- "nature" (corresponding to the woman's body, which appears on the scene "just as nature intended", that is to say, naked) *versus* "culture" (the codes regulating clothing, and even more strictly uniforms);
- "vulnerability" (at first glance, that of the protester's naked body) *versus* "invulnerability" (apparently ensured by the helmets and uniforms worn by the officers).

However, an interesting subversion intervenes on the last two aspects, as a result of the particular "practice", or use of nakedness made by the young woman: fully *aware* of her nakedness, the Naked Athena intentionally uses it to reveal—to "lay bare", one might provocatively say—the dysphoric connotation of the culture she is confronting—which highlights the inability of the opposition "nature" *versus* "culture" to properly describe this opposition.

Through such an "exercise of nudity", the protester is also able to escape the vulnerability generally attributed to the naked body. Hers is a *political* nudity, which underlines precisely the strength[12] of her naked body: as Tommaso Ariemma remarked, in fact, "if there is an ancestral strength in the human being, it is undoubtedly that of *taking charge of* being naked"[13] (in Gaspari 2020, my translation and emphasis).

This has interesting effects in terms of meaning-making processes, especially as related to the cognitive dimension and the interaction between the actants (or "cognitive subjects") involved in the scene—i.e. the *observer* ("the cognitive subject which is delegated by the enunciator and installed by it, through the process of disengagement, within the utterance-discourse", Greimas and Courtés 1979, Engl. Trans. 1982: s.v. "observer") and the *informant* (i.e. "a cognitive subject, endowed with a knowledge (partial or total) by the enunciator and installed by the latter, in the discourse, in a position of mediator vis-à-vis the enunciatee", *Ibid.*: s.v. "informant"). In fact, the naked body *wants to inform* the observer, which makes the latter (i.e. the police officers) *not able not to observe* it, in a regime of "exposure" (Fontanille 1989, see Fig. 6.1).

This radically alters the power relations involved in the situation: by means of her naked body, the informant (the Naked Athena) finds herself in a position of power (or—in Greimassian terms[14]—of "freedom", i.e. *being able to do*), despite the vulnerability (or "powerlessness", i.e. *not being able to do*) generally ascribed to the presumed naturality of nakedness. On the contrary, the observer (the police

[12] As the name chosen to refer to her, which makes clear reference to the ancient Greek goddess of war, handicraft and practical reason, also highlights.

[13] "Se c'è una forza ancestrale nell'essere umano è senza dubbio quella di farsi carico del fatto di essere nudi".

[14] See Greimas and Courtés (1979), Engl. Trans. 1982: s.v. "being-able (to do or to be)".

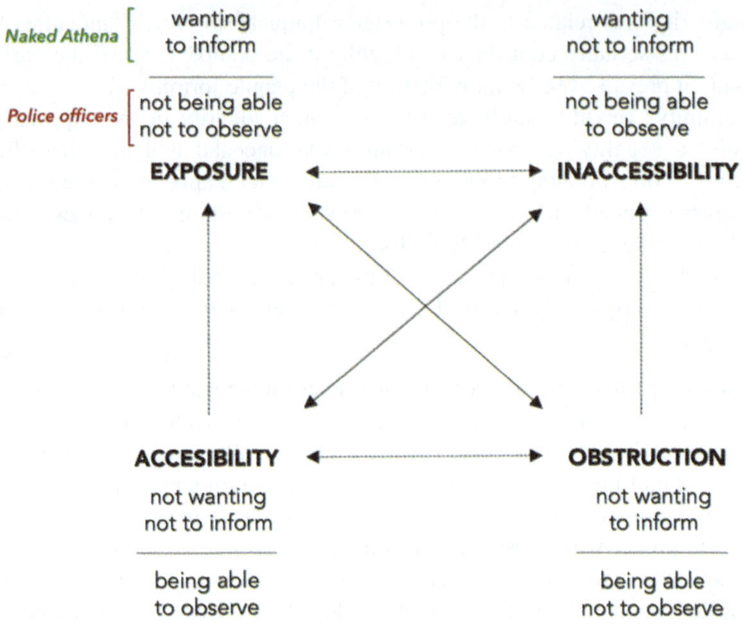


**Fig. 6.1** Semiotic square of the cognitive dynamics involving the Naked Athena and the police officers, based on the model introduced by Jacques Fontanille (1989)

"body"), whose authority is highlighted precisely by the uniforms covering it, finds itself in a position of "submission" or "obedience" (i.e. *not being able not to do*).

### 6.5.2 *Marina Abramović: Nudity, Performance Art and Agency*

As I highlighted above, it is particularly interesting to focus on the artistic representation of nudity, since it reflects—but sometimes can also contradict—sociocultural perceptions and inclinations towards the naked body. For the purposes of this work, it is even more interesting to focus not only on how and when the latter has been represented in the arts, but also and especially on how it can become the very "enunciator" of artworks. This is the case of performance art, that is to say, of "artworks that are created through actions performed by the artist or other participants, which may be live or recorded, spontaneous or scripted" (Tate Modern *n.d.*[15]).

One of the most renown performance artists in the world is Marina Abramović, who has elected precisely the body—and, more precisely, the naked body—as a

[15] https://www.tate.org.uk/art/art-terms/p/performance-art (Accessed: 10 October 2022).

fundamental element of her artistic reflection and activity. As Cristina Demaria observed,

> From the 1970s, when the so-called body art emerged as a visual genre, offering the artist's body as a naked site of inscription, up to the present, when performing has become a more playful and direct transmission of energy between the doer and the viewer, the work of Abramović represents an effective and powerful example of the body-as-a-text in which subjectivity can be re-expressed and reinvented through the transformations of the relation between time and space (Demaria 2004: 295).

In the six-hour performance *Rhythm 0* (Studio Morra Gallery, Naples, 1974), for instance, Abramović decided to assume a passive role, staying motionless and inviting the audience to do whatever they wished to her body, using in any way one of the 72 objects (including a rose, feather, perfume, honey, bread, grapes, wine, scissors, a scalpel, nails, a metal bar, a gun and a bullet, among others) placed on a table close to them.[16] At first the audience reacted with caution and modesty, but as time passed by, they became more and more intense and violent. By the end of the performance, the artist's body had been stripped, threatened and wounded, as she also commented later: "I felt really violated: they cut up my clothes, stuck rose thorns in my stomach, one person aimed the gun at my head, and another took it away. It created an aggressive atmosphere" (in Daneri et al. 2002: 20).[17]

After six hours, as planned, Abramović started walking towards the audience, returning to actively "inhabit" her body, undressed and wounded, and looking directly at the public, while "everyone ran away, to escape an actual confrontation" (*Ibidem*). In other terms, when the artist's body was "re-dressed", not with cloths, but with agency, taking charge of the nakedness and wounds imposed on it, a subversion similar to the one observed above occurred: leaving its initial position of "powerlessness" (*not being able to do*) and interdiction (*having not to do*), it reached the position of "permission" (*not having not to do*) and "liberty" (*being able to do*), precisely by "exposing" (*wanting to inform*) itself to the public, which therefore could not but observe it (cf. Fontanille 1989), becoming unable to confront her.

These same dynamics can be observed in another work by Marina Abramović, *Imponderabilia*, which she performed with Ulay in the Municipal Gallery of Modern Art in Bologna, Italy, in 1977. Standing naked in the main entrance of the museum, facing each other, the artists forced the public entering it to pass sideways through

---

[16] The artist's instructions reported: "There are 72 objects on the table that one can use on me as desired. Performance. I am the *object*. During this period I take full responsibility" (in Ward 2012: 19, my emphasis).

[17] A similar description was provided by art critic Thomas McEvilley: "It began tamely. Someone turned her around. Someone thrust her arms into the air. Someone touched her somewhat intimately. The Neapolitan night began to heat up. In the third hour all her clothes were cut from her with razor blades. In the fourth hour the same blades began to explore her skin. Her throat was slashed so someone could suck her blood. Various minor sexual assaults were carried out on her body. She was so committed to the piece that she would not have resisted rape or murder. Faced with her abdication of will, with its implied collapse of human psychology, a protective group began to define itself in the audience. When a loaded gun was thrust to Marina's head and her own finger was being worked around the trigger, a fight broke out between the audience factions" (in Ward 2012: 120).

the small space between their bodies and to choose which one of them to face (as Abramović herself clarifies in the opening of the compilation version of the performance registration). After this "living[, naked] door" (Abramović in Channell 2017), a text on the wall explained: "Imponderable. Such imponderable human factors as one's aesthetic sensitivity/the overriding importance of imponderables in determining human conduct", evidently recalling the visitors' behaviour (filmed by a hidden camera, which also became visible only at this very moment) on entry. Again, therefore, nudity was used to subvert the relation between vulnerability and power (as the artists themselves highlighted in their description of the work, cf. *Ibidem*), interdiction (i.e. *having not to do*), or prescription (i.e. *having to do*), and liberty.

This also remarks that, in fact, nudity as such is "unattainable" (cf. Agamben 2005): although more or less deprived of clothes, the body (when it is fully perceived as such, and not simply seen as an "inhabited object") necessarily remains "covered"—by a series of values related to gender, power (or powerlessness) and aesthetics, as well as to the multiple roles and positions that can be assumed with respect to it.

## 6.6 Concluding Remarks

In the first part of this chapter, I highlighted how different cultures and paradigms have alternatively euphoricised or stigmatised nudity, both as related to everyday life and collective representations (e.g. artworks, media discourses, etc.). In fact, as Ugo Volli remarked, "in this part of material culture, there are no constant or universal combinations … Contingency and history reign"[18] (Volli 1998: 5–6, my translation).

This makes nakedness clearly emerge as "not a being, nor a quality" (Ferrari and Nancy 2002: 31, my translation)—i.e. an inherent, or "natural" feature. "It is always a *relationship*, multiple *simultaneous relationships*, with others, with oneself, with the image, with the absence of image" (*Ibidem*, my translation and emphasis). From a semiotic point of view, in other terms, it is an effect of meaning, namely the outcome of a series of discursive operations—i.e. a cultural aspect through and through—that promote certain values, images, interpretations, while discouraging others.

This was also confirmed by the analysis of the case studies related to the use of nudity in protests and performance art, where specific processes of meaning-making and (re)valorisation emerged, as a result of an act of exposure. Thus, nakedness emerged as a powerful "political tool", which constantly questions the borders between nature and culture, the material level and the symbolic dimension, the biological sphere and the realm of semiosis. Not only our body, either covered in cloths or exposed in its nudity, is always "written" (cf. Volli 1998: 9), as a result of a series of practices (e.g. care, treatments, aesthetic standards, etc.) that intentionally or accidentally modify its appearance. But its exposure (i.e. its nakedness) can be

[18] "In questa parte della cultura materiale non esistono combinazioni costanti o universali …. Regnano la contingenza e la storia".

deliberately used to re-organise our relationship with other bodies, with the environment in which it is inserted (and with which it necessary interacts), and with the socio-cultural system of which it is part.

Far from simply representing a "natural state", therefore, nudity emerges precisely as one of the domains where it appears particularly urgent to dismiss the "nature/culture" dichotomy in view of a general rethinking of such categories, and of corporeality[19] itself.

## References

Abramović, Marina, Klaus Peter Biesenbach and MoMA. 2010. *Marina Abramović: The Artist Is Present*. New York, NY: Museum of Modern Art.

Agamben, Giorgio. 2005. Das verlorene paradiesische Kleid. Theologie der Nacktheit: Vanessa Beecrofts Berliner Performance. Frankfurter Allgemeine Zeitung, 84, April 12, 37.

Agamben, Giorgio. 2009. *Nudità*. Rome: Nottetempo.

Almeida, Dimitri. 2018. Marianne at the beach: The French burkini controversy and the shifting meanings of republican secularism. *Journal of Intercultural Studies* 39 (1): 20–34.

Ariemma, Tommaso. 2006. *Il nudo e l'animale*. Rome: Editori Riuniti Univ. Press.

Ariemma, Tommaso. 2007. *Il senso del nudo*. Milan-Udine: Mimesis.

Ariès, Philippe and Georges Duby, eds. 1987. *A History of Private Life. Vol. I: From Pagan Rome to Byzantium*. Cambridge, MA: Belknap Press of Harvard University Press.

Arieti, James A. 1975. Nudity in Greek athletics. *The Classical World* 68 (7): 431–436.

Augustine of Hippo. 413–426. *De civitate Dei (contra paganos)*. English Trans. Babcock, William, with notes by Ramsey, Boniface. 2012. *The City of God*. Hyde Park, NY: New City Press.

Bahrani, Zainab. 1996. The hellenization of Ishtar: Nudity, fetishism, and the production of cultural differentiation in ancient art. *Oxford Art Journal* 19 (2): 3–16.

Bonfante, Larissa. 1975. *Etruscan Dress*. Baltimore, MD, and London, UK: The Johns Hopkins University Press.

Bonfante, Larissa. 1989. Nudity as a costume in classical art. *American Journal of Archaelogy* 93 (4): 543–570.

Brown, Nadia E., and Sarah Allen Gershon, eds. 2020. *Body Politics*. Abingdon, UK, and New York, NY: Routledge.

Brownie, Barbara. 2017. *Acts of Undressing: Politics, Eroticism, and Discarded Clothing*. London, UK, and New York, NY: Bloomsbury Academic.

Burkert, Walter. 1979. *Structure and History in Greek Mythology and Ritual*. Berkeley-Los Angeles-London: University of California Press.

Chare, Nicholas, and Ersy Contogouris, eds. 2022. *On the Nude: Looking Anew at the Naked Body in Art*. New York, NY: Routledge.

Cicero, Marcu Tullius. c. 45 BCE. *Tusculanae disputationes*. English Trans. King, John Edward. 1927. *Tusculan Disputations*. Cambridge, MA: Harvard University Press.

Daneri, Anna, Giacinto Di Pietrantonio, Lorand Hegyi, Societas Raffaello Sanzio, and Angela Vettese, eds. 2002. *Marina Abramović*. Milan: Charta.

Demaria, Cristina. 2004. The performative body of Marina Abramović: Rerelating (in) time and space. *European Journal of Women's Studies* 11 (3): 295–307.

[19] Which, too, by no means can be seen as a "natural datum", since its material component cannot be disconnected from cultural dynamics and meaning-making processes, as it is becoming increasingly evident in contemporary cultures (see in particular Fuenmayor 2005; Stano 2019; Finol 2021).

Derrida, Jacques. 2006. *L'Animal que donc je suis*, ed. Marie-Luise Mallet. Paris: Éditions Galilée. English Trans. Wills, David. 2008. *The Animal That Therefore I Am*. New York, NY: Fordham University Press.
Descartes, René. 1637. *Discours sur la Méthode*. Leyde: Imprimerie de Ian Marie.
Deschênes, Stéphane. 2016. *The Official INF-FNI Definition of Naturism*, 1–6. https://downloads.inf-fni.org/download/51/english/133/definition-of-naturism-1974.pdf. Accessed 17 Nov 2022.
DuTertre, Jean-Baptiste. 1654. *Histoire générale des îles Saint-Christophe, de la Guadeloupe, de la Martinique et autres de l'Amérique*. Paris: Jacques et Emmanuel Langlois.
Fernández de Oviedo y Valdès, Gonzalo. 1526. *Sumario de la natural historia de las Indias*. http://bdh-rd.bne.es/viewer.vm?id=0000050339&page=1. Accessed 16 Nov 2022.
Ferrari, Federico and Jean-Luc Nancy. 2002. *Nus sommes: la peau des images*. Bruxelles: Gevaert.
Finol, José Enrique. 2021. *On the Corposphere. Anthroposemiotics of the Body*. Berlin: De Gruyter.
Fontanille, Jacques. 1989. *Les espaces subjectifs. Introduction à la sémiotique de l'observateur*. Paris: Hachette.
Fontanille, Jacques. 2004. *Soma et séma. Figures du corps*. Paris: Maisonneuve & Larose.
Fuenmayor, Victor. 2005. Entre cuerpo y semiosis: La corporeidad. *Opción* 21 (48): 121–154.
Gaspari, Ilaria. 2020. Il tabù della nudità da Adamo ed Eva a The Naked Athena. *Il Libraio.it*. https://www.illibraio.it/news/dautore/tabu-nudita-1387013/. Accessed 30 Nov 2022.
Greimas, Algirdas J., and Joseph Courtés. 1979. *Sémiotique: Dictionnaire raisonné de la théorie du langage*. Paris: Hachette. English Trans. Crist, Larry, Daniel Patte, James Lee, Edward McMahon II, Gary Phillips, and Michael Rengstorf. 1982. *Semiotics and Language: An Analytical Language*. Bloomington, IN: Indiana University Press.
Habinek, Thomas, and Alessandro Schiesaro. 1997. *The Roman Cultural Revolution*. Cambridge, UK: Cambridge University Press.
Hicks, Edward Lee. 1882. *A Manual of Greek Historical Inscriptions*. Oxford, UK: Clarendon Press.
Kyle, Donald G. 2015. *Sport and Spectacle in the Ancient World. Ancient Cultures*, 2nd ed. Chichester: John Wiley and Sons Inc.
Lu, Sheldon H. 2007. The naked body politic in postsocialist China and the Chinese diaspora. In *Chinese Modernity and Global Biopolitics: Studies in Literature and Visual Culture*, 71–92. Honolulu: University of Hawai'i Press.
Lunceford, Brett. 2012. *Naked Politics: Nudity, Political Action, and the Rhetoric of the Body*. Lanham, MD: Lexington Books.
Lunceford, Brett. 2019. *Public Nudity and the Rhetoric of the Body*. Lanham, MD: Lexington Books.
Mouratidis, John. 1985. The Origin of Nudity in Greek Athletics. *Journal of Sport History* 12 (3): 213–232.
Nguyen, Ryan. 2020. "Naked Athena": The story behind the surreal photos of Portland protester. *The Oregonian*, July 19 (updated July 21, 2020). https://www.oregonlive.com/portland/2020/07/the-story-behind-the-surreal-photos-of-portland-protester-naked-athena.html. Accessed 19 Nov 2022.
Plato. 1892. *The Dialogues of Plato in Five Volumes*, trans. Benjamin Jowett. New York, NY: Oxford University Press. https://oll.libertyfund.org/title/plato-the-dialogues-of-plato-in-5-vols-jowett-ed. Accessed 3 Nov 2022.
Pudor, Heinrich. 1906. *Katechismus der Nackt-Kultur*. Berlin-Steglitz: Verlag.
Salomon, Nanette. 1997. Making a world of difference: Gender, asymmetry, and the Greek nude. In *Naked Truths*, ed. Ann Olga Koloski-Ostrow and Claire L. Lyons, 197–219. London: Routledge.
Sorabella, Jean. 2008. The nude in the Middle Ages and the Renaissance. In *Heilbrunn Timeline of Art History*. New York, NY: The Metropolitan Museum of Art. http://www.metmuseum.org/toah/hd/numr/hd_numr.htm. Accessed 16 Nov 2022.
Stano, Simona. 2019. La soglia del senso. Il corpo come istanza semiotica. In *Il programma scientifico della semiotica*, ed. Massimo Leone, 149–162. Rome: Aracne.
Steinberg, Leo. 1983. *The Sexuality of Christ in Renaissance Art and in Modern Oblivion*. New York: Pantheon.

Ungewitter, Richard. 1906. *Die Nacktheit in entwicklungsge- schichtlicher, gesundheitlicher, moralischer und künstlerischer Betrachtung*. Stuttgart: Ungewitter.

Ungewitter, Richard. 1921 [1908]. *Nackt: eine kritische Studie*. Stuttgart: Selbstverlag.

Volli, Ugo. 1998. Una scrittura del corpo. In *Pelle di donna. Moda e bellezza*, ed. Adriana Moltedo, 1–45. Rome: Stampa Alternativa.

Ward, Frazer. 2012. *No Innocent Bystanders: Performance Art and Audience*. Hanover, NH: Dartmouth College Press.

Westcott, James. 2010. *When Marina Abramovic Dies: A Biography*. Cambridge, MA: The MIT Press.

Williams, James D. (ed.). 2009. *An Introduction to Classical Rhetoric: Essential Readings*. Chichester and Malden, MA: Wiley-Backwell.

## *Filmography:*

Channell, Louisiana. 2017. *Marina Abramovic & Ulay Interview: A Living Door of the Museum*. Louisiana Museum of Modern Art. United States, 4'19". https://channel.louisiana.dk/video/marina-abramovi%C4%87-ulay-living-door-museum. Accessed 23 Nov 2022.

# Chapter 7
# A Walk Through the Light(s) and the Path Towards an Internatural Turn

**Abstract** This chapter draws the conclusions of the research presented in this book, reconsidering the ambiguities and limits of the conception and representation of Nature in contemporary Western cultures, as related to the case studies analysed in the previous chapters. A new and particularly relevant case study is finally discussed in the attempt to move forward towards an "internatural turn" and a better comprehension of "natures"—now in the plural and, definitely, non-capitalised form.

## 7.1 "Hommage à la Nature": The 2022 Edition of the *Fête des Lumières* in Lyon

As I was writing the last part of this book, I saw an online post that immediately called my attention: the 2022 edition of the Festival of Lights (*Fête des Lumières*) in Lyon, France, had a section[1] specifically dedicated to nature!

In fact, the Festival's webpage and materials insisted on six main thematic routes ("6 univers thématiques différents"), including precisely the section *Hommage à la nature* ("A tribute to Mother Nature"[2]), in addition to: *A hauteur d'enfants* ("Family Friendly"), *Créations lyonnaises* ("Lyon's Got Talent"), *Regards d'ailleurs* ("Views from the Ouside"), *Grands spectacles* ("Larger than Life") and *Balade buissonnière* ("Off the Beaten Track"). While some categories were based on specific targets (e.g. *A hauteur d'enfants*) or experiences (e.g. *Grands spectacles*, *Balade buissonnière*), others mainly focused on the origin of the artists who created the installations (i.e. local artists—*Créations lyonnaises*—*vs.* external artists—*Regards d'ailleurs*), and only one was actually referred to a specific topic, which was precisely the one devoted to nature.

---

[1] *Hommage à la nature*, lit. "A homage to nature", translated as "A tribute to Mother Nature" on the Festival's official website, https://www.fetedeslumieres.lyon.fr/fr/selection/hommage-la-nature—English version: https://www.fetedeslumieres.lyon.fr/en/page/theme-selection (Accessed: 26 December 2022).

[2] All the English translations are taken from the official website, https://www.fetedeslumieres.lyon.fr/en/page/theme-selection (Accessed: 22 December 2022).



S. Stano, *Critique of Pure Nature*, Numanities - Arts and Humanities in Progress 26,
https://doi.org/10.1007/978-3-031-45075-4_7

**Fig. 7.1** Official poster of the *Fête des Lumières* 2022, Lyon, France (online version[3])

Indeed, such a focus was evident since the release of the official poster of the Festival (Fig. 7.1), which represented a prismatic composition made of light blue (on the sides) and white and yellow (in the centre) symmetric figures resembling butterflies, and at the same time recalling the form of semi-flowers, superimposed on each other, simulating a circular movement around the central writing "FÊTE DES LUMIÈRES LYON 2022" (with a vertical subdivision into two columns, reinforced by particular chromatic and eidetic configurations distinguishing the first three words, reported on the left in a bold white font, and the last two, reported on the right in thinner light blue and yellow letters), followed by the indication of the days and times of the Festival.

A video animation[4] also reinforced this figurative and thematic isotopy, by making such figurative patterns appear on the screen "fluttering" like butterflies and finally opening wide to form the final composition, with one of them landing on the initial letter of the word "Fête".

Hence, I decided to further explore this case, travelling to Lyon for a "walk through the Lights", which eventually proved essential for shedding "new light" on my research, revealing an original and very interesting perspective on the discursivisation of nature. This chapter examines such a perspective (especially in Sect. 7.3), also showing how it relates to a possible path towards an "internatural turn". Before doing so, however, it briefly reconsiders the case studies analysed in the previous chapters, highlighting their main implications and problematic aspects (Sect. 7.2).

---

[3] https://www.fetedeslumieres.lyon.fr (Accessed: 22 December 2022).

[4] Now available as part of the video "Retour sur la Fête des Lumières 2022 !" on the Festival's main page, as well as on the YouTube channel of the city of Lyon (Ville de Lyon 2022), https://www.youtube.com/watch?v=BrlJMJJ0TpQ (Accessed: 26 December 2022).

## 7.2 A Recurring Dichotomy: Isotopies, Limits, and Problems of Today's Praised Nature

After an initial problematisation of the concept of nature and its evolution over time (Chap. 2), the focus of attention of the research moved to the discursivisation of such a concept in various domains, in the aim to highlight the main figures and effects of meanings arising from it.

More specifically, Chap. 3 investigated the collective imaginary of Mother Nature, from classical iconography to contemporary popular culture (especially as related to music, cinema, television, and advertising). A series of interesting case studies highlighted the transition of such a figure from a divine connotation (typical of ancient imaginaries) to an increasingly more "secularised" and sometimes even objectified characterisation. This, as the chapter showed, has had an evident impact on the roles and values attributed to Mother Nature: no longer seen as a creator of the world and its inhabitants, or a nurturing principle (as the changes in the representation of her body also highlighted), in modern and contemporary representations she has progressively been depicted as a simple "supervisor" of the natural order, which simply ensures its respect and proper development. Consequently, her figurativisation has lost most elements symbolising extra-ordinary abilities (e.g. fairy wings or wands, which only remain in comics and TV shows mainly addressed to children, and generally with limited powers), to include at most particular suits or ornaments highlighting her—not always absolute, nor entirely effective, as I showed through various examples—leadership. What is more, a voyeuristic look also seems to have been projected on Mother Nature on some occasions, "unveiling" her body until laying it bare, thus completely objectifying her. In some cases, the loss of all anthropomorphic features was also observed as a way to stress the separation and opposition between nature and culture—especially when Mother Nature was presented as dead or dying, precisely as a result of the action of man (i.e. the culture to which it is opposed) and its "catastrophic" consequences. This also pointed to a more demanding, intimidating and self-centred figure, which "doesn't need people" (while people need her). As a counterpart to such a "dramatizing" trend insisting on the loss or estrangement of Mother Nature, the chapter also highlighted its "ironicisation", especially in advertising. However, a dualistic opposition between nature and culture remained dominant, with any attempt to combine these two dimensions being depicted as "naïve" or inevitably destined to fail—either by means of a progressive separation or through an overcoming, which sometimes was also represented as a real annihilation.

Drawing on the aspects outlined in relation to the figure of the "dying (Mother) Nature", Chap. 4 focused on the analysis of the collective imaginaries of catastrophic events, ranging from fictional discourses (with a particular focus on cinema) to non-fictional representations (especially as related to the Anthropocene, environmental discourses and the Covid-19 pandemic). The discussion of relevant case studies pointed out the interconnection between "natural" disasters or calamities, on the one hand, and "cultural" catastrophes, on the other hand, highlighting—in accordance with relevant literature in the field—that, while the former seem to be at least partially related to ecological or social dislocations caused by (sometimes unconscious, sometimes irresponsible) human actions, the latter can be envisaged as natural entropic degenerations of complex organisations. Hence, the chapter introduced a general model, or "matrix", encompassing both a process of *naturalisation* of nature, which sees it as completely independent from human action, and a process of *culturalisation* of nature, which rather relies on the idea of an excessive and reckless exploitation of natural resources by a dysphoric cultural system (i.e. the one generally "overturned" by the catastrophic event). This was further discussed considering a more recent case study, *Don't Look Up* (McKay 2021), which interestingly shows how nature is intrinsically linked to culture, as it is inevitably experienced and conceived through it. However, as various non-fictional discourses demonstrated, a dichotomic view of the relation between nature and culture seems far from be abandoned. While being celebrated by some as the collapse of all dichotomic views opposing Nature to Culture, the widespread idea of Anthropocene, for instance, has mainly found expression in narratives that further accentuate such an opposition, contrasting natural resources, on the one hand, and mankind (either in the form of humanity as a whole or as related to specific groups or socio-economic systems, such as capitalism, etc.), on the other hand. Accordingly, a number of discourses focusing on environmental issues maintain a dichotomic view, representing nature as a reality ontologically different and separate from the socio-cultural dimension, which has reached a "critical point" precisely because of human activities. What is more, as I showed, this ideology is recurrent also in other domains, as exemplified by a series of texts referred to the Coronavirus pandemic. Conceiving the Covid-19 outbreak as a "catastrophe" marking the end of a dysphorised culture thought to have imposed itself over nature, in fact, such texts have further enhanced a process of idealisation of nature, stereotypically depicting it as an original, idyllic and somehow "sacred" state, as opposed to the dysphoric "culture" of current societies. Moreover, unlike fictional representations, they have largely favoured a denial of culture as such—i.e. a sort of annihilation of man himself (or, at least, of his activities) in view of a "re-naturalisation" of our planet—over the passage to alternative cultural systems. Once again, and even more than in fictional collective imaginaries, then, nature is conceived as an alterity, namely as something that is not only separate from, but also opposite to culture, perpetrating a dualistic vision that fails to recognise man himself as part of the natural world.

Chapter 5 then reconsidered the opposition between nature and culture in the food domain, where it has always played a crucial role. While the transition from the former to the latter was euphoricised in ancient societies, with cooking representing a general metaphor of all vital processes, in the 17th century a preference for "fermentation" (i.e. a more "natural" transformation of food) arose. What is more, a marked "uncooking" trend, which seems to go well beyond the praise of foods not requiring elaborated cooking procedures to embrace a radical rejection of any cultural transformation of food, was observed in contemporary foodspheres. In order to better describe this phenomenon, relevant case studies related to "clean eating",

GMOs *vs.* organic foods, the Mediterranean diet, and the aphrodisiac potential traditionally ascribed to certain food products were analysed through a semiocultural approach. In all these cases, a marked process of dysphorisation of culture emerged: no longer conceived as an "order", as it was in the culinary systems studied by Lévi-Strauss, it rather tends to be depicted as an "insub-ordination" to the laws of nature. Today's "myth of the natural" therefore manifests itself through a strong opposition to any modification of food, from refinement to industrialisation, from genetic alteration to new forms of production. In some cases, moreover, culture seems totally neglected: the analysed representations of aphrodisiac foods, for instance, outlined a tendency to attribute some sort of "alchemical potential" to substances themselves, disregarding the fact that they are always and necessarily part of specific practices of preparation and consumption, which actively and incessantly re-define their meanings, and hence also their possible effects on those who eat them. If any cultural trait is admissible, as the case of the Mediterranean diet proved, it is at most through a process of naturalisation, which "crystallises" it into a never-changing "tradition", closed to transformation and innovation. Moreover, even in this case, the tension with culture (which generally takes on the forms of modernity and globalisation) remains dominant, further remarking a differential conception of nature as its opposite.

Such a tension also proved crucial in the dynamics analysed in the following chapter. Building on the importance of the corporeal dimension and its "unveiling" (as described in Chap. 3), Chap. 6 focused on the analysis of the naked body between its supposed natural ascription and its multiple cultural characterisations. While nudity is mostly associated with nature in common sense, various meanings have been attributed on it over time, both as related to everyday life and collective representations—as the chapter showed through a diachronic overview ranging from ancient times to the present days. This was also confirmed by the analysis of some relevant case studies related to the use of nudity in contemporary protests and performance art, where specific processes of meaning-making and (re)valorisation were identified precisely as a result of "exposure". Nakedness therefore emerged as a powerful tool in political and social action, specifically because it is able to constantly renegotiate the borders between nature and culture, the material level and the symbolic dimension, the biological sphere and the realm of semiosis. Once again, in other terms, the urgency to dismiss the "nature/culture" dichotomy in view of a general rethinking of such categories arose, confirming what examined in the previous chapters.

Despite their peculiarities, in sum, all the cases considered in the previous chapters reveal a specific ideology of Nature, which is essentially based on a differential conception that, interpreting it as the "opposite" of culture, results in great indeterminateness, leading—as it was highlighted for each case study—to a number of inconsistencies and contradictions. What is more, such an ideology relies on a rather simplistic and unrealistic view, which is unable to account for the fact that—as Chap. 2 illustrated in detail—nature is itself necessarily an artefact, the result of discursive practices, an effect of meaning—and that, as such, it is part of culture as much as culture is part of it.

## 7.3 Seeing the Light(s): A Possible Path Towards an Internatural Turn

Encountered while working on the final part of this book, the discourses figurativising nature in the context of the *Fête des Lumières* 2022 proved particularly thought-provoking, as they evidently differed from what observed in the other case studies. The following paragraph focuses on their analysis, leading to the final section, where I will present some concluding remarks on such interesting case studies, as well as more general reflections on the potential of artistic creativity—and semiotic research—in tracing a possible path towards an internatural turn.

### *7.3.1 Nature(s) at the* Fête des Lumières *2022*

In the above-described poster of the event (Fig. 7.1), the references to the natural world found expression in markedly stylised and digitalised figures, which rather than simply imitating natural forms and colours, seemed to markedly hybridise them with the visual marks of geometry (i.e. well-defined shapes, characterised by clear contours and a perfect symmetry) and technology (i.e. very vivid, markedly artificial, colours; the prismatic configuration of the resulting image, which recalls the phenomenon of refraction of light, namely a "natural" element reinterpreted in a technological—that is, "cultural"—key by the Festival), creating an original synthesis of nature and culture that challenged the dualistic vision identified in the texts analysed in the previous chapters.

Such a synthesis also characterised various art installations included in the thematic section "Hommage à la nature", whose official description insisted precisely on the "particular relation" they had with nature ("Les œuvres rassemblées ici ont une *relation toute particulière* à la nature et à son cadre enchanteur", my emphasis), further specified in terms of an "inspiration" and "celebration" in the English version ("The works grouped under this theme take *inspiration* from and *celebrate* the wonders of nature", my emphasis). Such ideas, however, did not relate to the image of a deity distant and separate from man, but rather to the idea of an entity closely related to the social and cultural sphere.

*Le Parc du Nombre D'or*[5] by Javier Riera (Fig. 7.2), for instance, went beyond a dichotomic view of nature and culture by explicitly inter-connecting them, as remarked by the very presentation of the work:

> It all begins with a mathematical programme, illustrations, equations, golden ratios, and numerical data on nature. You could call this Javier Riera's palette! This artist loves to work with nature, in a completely innocuous way. He works with his materials like a painter, searching for parallels between the irregular nature of the landscape and the trees and the precision of the geometric forms that emerge from his data. For Riera, geometry is a language

[5] https://www.fetedeslumieres.lyon.fr/fr/oeuvre/le-parc-du-nombre-dor (Accessed: 26 December 2022).

**Fig. 7.2** *Le Parc du Nombre D'or*, Javier Riera, Parc de la Tête d'Or, Lyon, France (*Fête des Lumières* 2022)

> that provides a bridge between nature and our innermost selves. When we watch this show and enter into his world, we are all invited to discover the subtle facets of nature that communicate and thrive within us. A wonderful experience in Lyon's most emblematic park.

Nature emerged in this case as both the artist's canvas, namely the surface on which to realise the artwork, and his palette, that is to say, the source of the data used to realise it.

By virtue of the culture he represents (within which technology plays a crucial role), Riera created his artistic installation not only by "translating" the latter into the "precision" (i.e. regularity) of geometric figures, but also by projecting such figures on the (irregular) foliage of trees and bushes, thus making it match such regularity.

The same elements were at work also in Onionlab's *Agorythm*,[6] which used a particular "algorithm" for the elaboration of Lyon's environmental data and its transposition into the "rhythmed" succession of laser beams and light sculptures. These were finally projected onto a central common space (i.e. an "agora"), favouring the visitors' immersion into the artwork itself (Fig. 7.3).

A technological "translation" was also present in *Urban Oracle*,[7] located at Allée Achille Lignon. Based on a system of Artificial Intelligence trained by the artistic studio Ultravioletto, this installation "dialogue[d] with the outside world" (as its description reported) in a divination performance, in the aim to erase all boundaries

[6] https://www.fetedeslumieres.lyon.fr/fr/oeuvre/agorythm (Accessed: 26 December 2022).

[7] https://www.fetedeslumieres.lyon.fr/fr/oeuvre/urban-oracle (Accessed: 26 December 2022).

**Fig. 7.3** *Agorythm*, Onionlan, Parc de la Tête d'Or, picture ©Laurence Danière, Lyon, France (*Fête des Lumières* 2022)

between the natural world (i.e. environment, setting, events) and the cultural universe (i.e. language, pre-figuration, symbolism).

A similar idea was at the base of *Alcove LTD*[8] (Encor Studio, Berges du Rhône), a long transparent shipping container to be walked around and observed from all angles:

> The Swiss designers use[d] changing lights, shapes and materials to explore the concept of isolation and changing spaces. … Liquid crystal film [wa]s used to transform the glass walls inviting us to enjoy an immersive experience where mirroring effects blur the lines. Facing this immobile container, almost communicating with it, our perception of the environment step[ped] outside the box.

As the description of the work remarked, the relation between nature and culture found here expression in the link between its interior and its exterior, with a mirroring effect used purposedly to make all borders "blur", and communication (intended in its etymological sense of "sharing", "making common") prevail over separation and distinction.

The same idea could be also found in another installation, which was not included in this thematic session, but nonetheless seemed to feature the same aspects: *Mirror Mountain*[9] by Nicolas Ticot (Fig. 7.4), placed at Place Bellecour.

A mountain made up of mirrors—as it can be inferred from the title—combined with light and sound, which could be activated by visitors themselves, offered a multisensorial experience, inviting people to adopt a new perspective on nature and culture, as the description of the work interestingly suggested: "What if we decided to look at ourselves and the world in a new light?".

---

[8] https://www.fetedeslumieres.lyon.fr/fr/oeuvre/alcove-ltd (Accessed: 26 December 2022).

[9] https://www.fetedeslumieres.lyon.fr/fr/oeuvre/mirror-mountain?edition=970 (Accessed: 26 December 2022).

**Fig. 7.4** *Mirror Mountain*, Nicolas Ticot, Place Bellecour, Lyon, France (*Fête des Lumières* 2022)

In addition to these particularly avant-gardist experiments, the installations more simply using technology to "mimic" nature also markedly insisted on its relation with culture, either by emphasising the complementarity of such dimensions (e.g. Studio Toer's *Firefly Field*,[10] which was made up of several flying light points imitating how fireflies move and react to the surrounding environment and agents) or by placing the human figure at the very centre of the performance (e.g. *Les Lumignons du Coeur*,[11] where spectators were invited to contribute to the ever-changing set of flames run by a team of volunteers, whose presence was therefore constant among the several re-creations of flowers and plants, as shown in Fig. 7.5).

The eco-friendly project *Gazouillis*[12] by BIBI pushed interaction further by involving the population in the very creation of the artwork. A giant aviary placed at Place Voltaire contained 300 luminous birds made of recycled plastic, which was collected through an appeal for plastic bottles launched before the Festival, and then crafted by local young people during workshops run by the artist himself. Direct action therefore substituted simple vision in this case, as it also occurred in the design and implementation of the installation *Beacon*[13] (by Craig Morrison and Emilien Guesnard, placed at Jardin de l'Institut Lumière), reinforcing the idea of

[10] https://www.fetedeslumieres.lyon.fr/fr/oeuvre/firefly-field (Accessed: 26 December 2022).

[11] https://www.fetedeslumieres.lyon.fr/fr/oeuvre/les-lumignons-du-coeur-2 (Accessed: 26 December 2022).

[12] https://www.fetedeslumieres.lyon.fr/fr/oeuvre/gazouillis (Accessed: 26 December 2022).

[13] https://www.fetedeslumieres.lyon.fr/fr/oeuvre/beacon (Accessed: 26 December 2022).

**Fig. 7.5** *Les Lumignons du Coeur*, Direction des Événements et de l'Animation de la Ville de Lyon, Parc de la Tête d'Or, picture ©Muriel Chaulet, Lyon, France (*Fête des Lumières* 2022)

a positive relation between nature and culture (which in this case was particularly marked by the value of sustainability).

Finally, in the project *Cachés dans la ville*[14] by Theoriz (Parc Blandan, Fig. 7.6, top), interaction was favoured by means of both visual and sound immersion, with digital technology blending with performance art to stimulate the spectators' senses and directly involve them in the varied set of offered experiences. In one of them, for instance, people were invited to use their own bodies and movements to direct lights and shadows towards different surfaces (and sometimes even to themselves), thus simulating the presence and movement of fish, birds and other natural species, and becoming an essential part of immersive practices set in the ocean, the space and other environments. Thus, the installation emerged as the counterpart of another work, external to the thematic section "Hommage à la nature"—i.e. *J'ai attrapé un...*,[15] by Stéphane Masson (Fig. 7.6, bottom)—, which also included projections of natural figures (e.g. a flying fish, a mini elephant, a solitary cloud), but "caught" them in a cage erected in a street in the Grolée District.

In *Cachés dans la ville*, not only the cage disappeared, but the relation between the inside and the outside was subverted, and the surrounding environment was integrated into the artwork, in order to favour a direct relation (i.e. an "immersion") between nature and culture.

---

[14] https://www.fetedeslumieres.lyon.fr/fr/oeuvre/caches-dans-la-ville (Accessed: 26 December 2022).

[15] https://www.fetedeslumieres.lyon.fr/fr/oeuvre/jai-attrape-un (Accessed: 26 December 2022).

**Fig. 7.6** *Cachés dans la ville*, Theoriz, Parc Blandan, picture ©Emmanuel Foudrot (top), and *J'ai attrapé un...*, Stéphane Masson (bottom), Lyon, France (*Fête des Lumières* 2022)

## 7.3.2 From the Lights to the Light: Concluding Remarks

While it is certainly early to recognise a new trend in the examples considered above, their novelty and variety should be highlighted, especially in relation to the possibility of tracing a path towards an "internatural turn", intended as the awareness—even at the level of shared imaginaries and communication—that nature is not at all an original, single entity separate from, and opposite to the socio-cultural dimension and its multiple variations, but rather a multiple and itself variable dimension that emerges precisely from its inter-action with culture(s), as well exemplified by the mixture of forms, rhythms and languages characterising both the poster of the Festival and the analysed artistic installations. In fact, if the processes of imagination and discursivisation of nature tend to rely on shared imaginaries "clothing" such a concept with specific figures (see Leone 2012, 2016; cf. Descola 2006, 2010), the examples examined in this chapter point out a relevant change in the contemporary Western perception of nature. No longer conceived as an ontological "niche" defined by the absence of humanity, it is rather manifestly related to it, by virtue of an "osmotic" process: nature and culture are shown as mutually influencing each other, combining regularity and irregularity, prefiguration and unpredictability, environmental elements and human bodies and actions.

This evidently recalls the idea of "multinaturalism" introduced by scholars such as Descola (2005) and Viveiros de Castro (2009), also expanding it in view of the processes described by Marrone (2011). In fact, the case studies considered in the previous pages effectively show that various "ontologies of the natural" are incessantly re-defined by the social discourses that animate the space of cultures. This is precisely what seems to open a path towards a regime of "internaturality", where natures—now in the plural and, definitely, non-capitalised form—are not only multiple, but also necessarily intertwined with each other, as well as with cultures, without any pre-established identity, but rather with incessant processes of "translation"—entirely comparable to the ones concerning cultural systems (cf. Lotman 1984, 1992)—continuously redefining them.

This is particularly relevant, as it marks an evident discrepancy from most contemporary discourses and representations of nature (such as the various examples considered in the previous chapters). What is more, it contributes to point out the potential of artistic creativity in opening the way to new codes, visions and interpretative paths. In fact, as research has shown (see in particular Lotman 1998; Ferraro 2012: 77), in the artistic domain the rule tends to be the result of, rather than the premise for the production of messages. Codes, in other terms, do not appear determinedly assumed or prefixed, but can be treated as fluid, adaptable and dynamic. As Jurij M. Lotman remarked, with respect to the norm, the code, and even the author's invention itself, the artistic text always appears as "something subjected to *unpredictable deviations*"[16] (Lotman 1998: 441, my translation and emphasis). Rather than simply "representing" their own culture in their works, therefore, artists can—and often actually do—distance themselves from it, reflect on it, and even challenge it, thus conferring a *meta-cultural* function on their creations (cf. Ferraro 2012: 153). To a certain extent already visible in the analysis of performance art as related to the body and nakedness (see Chap. 6), this aspect imposes itself more evidently in the cases analysed above, which manifestly recall a reflection on nature and culture, and particularly on the relationship of mutual dependence and interaction between them.

In this sense, art emerges as a real "modelling system" (cf. Lotman 1970), within which *conflictual relations* play a crucial role: artistic creativity makes values and

[16] "Qualcosa di sottoposto a imprevedibili deviazioni".

signs pertaining to different codes confront each other, causing a real "clash" of ideologies, that is to say, a "structural transformation of the meaning of the border between worlds of signs closed in themselves"[17] (Burini in Lotman 1998: 144, my translation). Thus, *programming*[18] (i.e. regularity) and *manipulation* (i.e. intentionality) leave the way to *accident* (i.e. chance) and *adjustment* (i.e. uncertainty, progressive change and adaptation), prefiguring new forms of agentivity and the possibility of innovative, and unexpected, interpretative paths.

An excellent field of investigation therefore opens up for any "critique" aiming at investigating nature(s)—or, better, as the analysis of the considered case studies has pointed out, the "mechanisms of naturality". In this sense, as the research has shown, semiotics can certainly play a crucial role. Leaving aside any attempt to "unveil" nature—which characterised the Science celebrated in the frontispiece of Blasius' *Anatome animalium* (1681), or the Philosophy praised in the opening pages of Peyrard's *De la Nature et des ses Lois* (1793)[19]—, in fact, the study of "the life of signs within society"[20] offers very effective tools for delving into the meshes of the veil the covers it. A veil that, rather than simply "hiding" nature, also *connects* us to it—as the previous pages hope to have effectively illustrated—, in a dense twine of meaning and matter, discourse and perception, language and reality, which demands us to be carefully scrutinized, rather than simply removed.

## References

Burini, Silvia. 1998. Postfazione. Jurij Lotman e la semiotica delle arti figurative. In *Il girotondo delle muse. Saggi sulla semiotica delle arti e della rappresentazione*, ed. Jurij M. Lotman, Silvia Burini, 131–164. Bergamo: Moretti&Vitali.

Descola, Philippe. 2005. *Par-delà nature et culture*. Paris: Gallimard. English Trans. Lloyd, Janet. 2013. *Beyond Nature and Culture*. Chicago, IL: The University of Chicago Press.

Descola, Philippe. 2006. La fabrique des images. *Anthropologie Et Sociétés* 30 (3): 167–182.

Descola, Philippe. 2010. *La fabrique des images: Visions du monde et formes de la représentation*. Paris: Musée du quai Branly-Somogy.

Ferraro, Guido. 2012. *Fondamenti di teoria sociosemiotica. La visione "neoclassica"*. Rome: Aracne.

Landowski, Eric. 2005. *Les interactions risquées—Nouveaux Actes Sémiotiques*, 101–103. Limoges: PULIM.

Leone, Massimo. 2012. Nasi possibili: Internaturalità e figurazione. In *Semiotica della natura (Natura della semiotica)*, ed. Gianfranco Marrone, 183–207. Milan-Udine: Mimesis.

Leone, Massimo. 2016. Nature and culture in visual communication: Japanese variations on Ludus Naturae. *Semiotica* 213: 213–245.

---

[17] "Un rinnovamento strutturale del senso del confine tra mondi di segni in sé chiusi".

[18] I adopt here the terminology introduced by Eric Landowski in *Les interactions risquées* (2005), where he focuses on interactional practices based on the processes of construction of meaning in everyday experiences. For further details, refer to the explanation of this model reported in Chap. 3 (see *supra*, Sect. 3.3).

[19] The reference is to the images analysed in Chap. 3 (see *supra*, Sect. 3.2).

[20] I.e. semiotics, as defined by de Saussure (1916; Engl. Trans. 1959: 16).

Lotman, Jurij M. 1970. *Struktura chudožestvennogo teksta.* Moscow: Iskusstvo. Italian Trans. Bazzarelli, E. 1972. *La struttura del testo poetico.* Milan: Mursia.
Lotman, Jurij M. 1984. О семиосфере. *Труды по знаковым системам (Signs Systems Studies)* 17: 5–23. English Trans. Clark, Wilma. 2005. On the semiosphere. *Signs Systems Studies* 33 (1): 205–229.
Lotman, Jurij M. 1992. *Kul'tura i vzryv.* Moscow: Gnozis. English Trans. Clark, Wilma. 2009. *Culture and Explosion.* The Hague: Mouton de Gruyter.
Lotman, Jurij M. 1998. *Il girotondo delle muse. Saggi sulla semiotica delle arti e della rappresentazione*, ed. Silvia Burini. Bergamo: Moretti&Vitali.
Marrone, Gianfranco. 2011. *Addio alla natura.* Torino: Einaudi.
de Saussure, Ferdinand. 1916. *Cours de linguistique générale.* Paris: Payot. English Trans. Baskin, Wade. 1959. *Course in General Linguistics.* New York: Philosophical Library.
Viveiros de Castro, Eduardo. 2009. *Métaphysiques cannibales.* Paris: PUF. English Trans. Skafish, Peter. 2014. *Cannibal Metaphysics: For a Post-Structural Anthropology.* Minneapolis, MN: Univocal.

## *Filmography*

McKay, Adam. 2021. *Don't Look Up.* United States, 138'.
Ville de Lyon. 2022. *Retour sur la Fête des Lumières 2022 !.* France, 4' 52". https://www.youtube.com/watch?v=BrlJMJJ0TpQ. Accessed 26 Dec 2022.

The manufacturer's authorised representative in the EU is Springer Nature Customer Service Centre GmbH, Europaplatz 3, 69115 Heidelberg, Germany. If you have any concerns regarding our products, please contact ProductSafety@springernature.com

Printed and bound by CPI Group (UK) Ltd, Croydon, CR0 4YY
28/09/2026
02224075-0017